A-LEVEL YEAR 2

STUDENT GUIDE

AQA

Chemistry

Inorganic and organic chemistry 2

Alyn G. McFarland

Nora Henry

HODDER
EDUCATION
AN HACHETTE UK COMPANY

Hodder Education, an Hachette UK company, Blenheim Court, George Street, Banbury, Oxfordshire OX16 5BH

Orders

Bookpoint Ltd, 130 Park Drive, Milton Park, Abingdon, Oxfordshire OX14 4SE

tel: 01235 827827

fax: 01235 400401

e-mail: education@bookpoint.co.uk

Lines are open 9.00 a.m.–5.00 p.m., Monday to Saturday, with a 24-hour message answering service. You can also order through the Hodder Education website: www.hoddereducation.co.uk

ISBN 978-1-4718-5863-5

First printed 2016

Impression number   5 4 3 2 1

Year   2020 2019 2018 2017 2016

This guide has been written specifically to support students preparing for the AQA A-level Chemistry examinations. The content has been neither approved nor endorsed by AQA and remains the sole responsibility of the authors.

Cover photo: Ingo Bartussek/Fotolia

Typeset by Integra Software Services Pvt. Ltd, Pondicherry, India

Printed in Italy by Printer Trento S.r.l.

Hachette UK's policy is to use papers that are natural, renewable and recyclable products and made from wood grown in sustainable forests. The logging and manufacturing processes are expected to conform to the environmental regulations of the country of origin.

Contents

Content Guidance

Questions & Answers

■ Getting the most from this book

Exam-style questions

Sample student answers

Practise the questions, then look at the student answers that follow.

Commentary on sample student answers

Read the comments (preceded by the icon **e**) showing how many marks each answer would be awarded in the exam and exactly where marks are gained or lost.

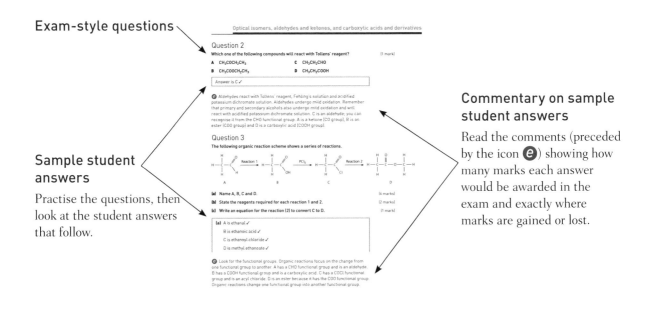

■ About this book

This book will guide you through sections 3.2.4 to 3.2.6 (inorganic chemistry) and 3.3.7 to 3.3.16 (organic chemistry) of the AQA A-level Chemistry specification. The content of the AS specification is examined in the AS examination and in the A-level examinations.

Paper 1 of A-level covers physical chemistry (3.1.1 to 3.1.12, found in the first and third student guides of this series), except 3.1.5 Kinetics and 3.1.9 Rate equations, as well as all inorganic chemistry (3.2.1 to 3.2.6, which can be found in the second student guide in this series and this book).

Paper 2 of A-level covers organic chemistry (3.3.1 to 3.3.16, found in the second student guide in this series and this book) as well as 3.1.2 to 3.1.6 and 3.1.9, which are covered in the first and third student guides of this series.

Paper 3 covers all content.

This book has two sections:
- The **Content Guidance** covers the A-level inorganic chemistry sections 3.2.4 to 3.2.6 and organic chemistry sections 3.3.7 to 3.3.16, and includes tips on how to approach revision and improve exam technique. Do not skim over these tips as they provide important guidance. There are also knowledge check questions throughout this section, with answers at the back of the book. At the end of each section there is a summary of the key points covered. Many topics in first-year sections, covered in the second student guide of this series, form the basis of synoptic questions in A-level papers. There are three required practicals related to the topics in this book and notes to highlight these are included.
- The **Questions & Answers** section gives sample examination questions (including synoptic questions) on each topic, as well as worked answers and comments on the common pitfalls to avoid. This section contains many different examples of questions, but you should also refer to past papers, which are available online.

The Content Guidance and Questions & Answers section are divided into the topics outlined by the AQA A-level specification.

Content Guidance

■ Inorganic chemistry

Properties of period 3 elements and their oxides

This section examines the properties of the elements in period 3 (sodium to sulfur) and their oxides, to include Na_2O, MgO, Al_2O_3, SiO_2, P_4O_{10}, SO_2 and SO_3.

Sodium

Sodium burns in air with a yellow flame, producing a white solid — sodium oxide:

$$4Na + O_2 \rightarrow 2Na_2O$$

Heat is released.

Sodium oxide is an **ionic**, basic oxide. Its melting point is 1275°C due to the strong forces of attraction between the oppositely charged ions, which require a lot of energy to break.

Sodium oxide reacts with water to produce a colourless solution of sodium hydroxide:

$$Na_2O + H_2O \rightarrow 2NaOH$$

This solution will have a pH of around 12–14.

Magnesium

Magnesium burns in air, producing a bright white light, releasing heat and forming a white solid — magnesium oxide.

$$2Mg + O_2 \rightarrow 2MgO$$

Magnesium oxide is an **ionic**, **basic oxide**. Its melting point is 2852°C because it has a strong lattice, owing to the small, highly charged Mg^{2+} and O^{2-} ions. There are strong forces of attraction between these small 2+ and 2− ions.

10% of the solid formed on igniting magnesium is magnesium nitride, Mg_3N_2.

Magnesium oxide is virtually insoluble in water but some does react to form magnesium hydroxide:

$$MgO + H_2O \rightarrow Mg(OH)_2$$

The pH of a solution is around 9, owing to the low solubility of magnesium oxide in water.

A basic oxide reacts with an acid, forming a salt.

Exam tip

The oxide ion is acting as a Brønsted–Lowry base. It accepts a proton from water, forming hydroxide ions:

$$O^{2-} + H^+ \rightarrow OH^-$$

The pH of the solution formed depends on the mass of sodium oxide used and the volume of water to which it is added. This could be a synoptic calculation.

A Brønsted–Lowry base accepts a proton.

Knowledge check 1

Explain why the melting point of MgO is greater than that of Na_2O.

Aluminium

Aluminium powder burns in air with a white light, producing a white solid — aluminium oxide:

$$4Al + 3O_2 \rightarrow 2Al_2O_3$$

Its melting point is 2070°C owing to the charges on the ions and the small size of the ions. A lot of energy is required to break the strong forces of attraction between the ions.

Aluminium oxide does not react with water. When mixed with water the pH remains at 7 because none of the Al_2O_3 dissolves or reacts.

Aluminium oxide is an **ionic**, amphoteric oxide. Amphoteric oxides react with both acids and bases (alkalis).

With acid: $Al_2O_3 + 6H^+ \rightarrow 2Al^{3+} + 3H_2O$

With alkali: $Al_2O_3 + 2OH^- + 3H_2O \rightarrow 2Al(OH)_4^-$

When aluminium oxide reacts with alkali, the aluminate ion, $Al(OH)_4^-$, is formed.

Overall equations can be written for the reactions with acids and bases:

$$Al_2O_3 + 6HCl \rightarrow 2AlCl_3 + 3H_2O$$

$$Al_2O_3 + 2NaOH + 3H_2O \rightarrow 2NaAl(OH)_4$$

$NaAl(OH)_4$ is sodium aluminate(III).

Silicon

Silicon burns in air, if heated strongly enough, to form silicon dioxide:

$$Si + O_2 \rightarrow SiO_2$$

Silicon dioxide is a **macromolecular (giant covalent)**, acidic oxide. Its melting point is 1610°C because a lot of energy is required to break the many strong covalent bonds in its structure. Silicon dioxide does not react with water because the water cannot break up the giant covalent structure.

Silicon dioxide is an acidic oxide, which reacts with alkalis:

$$SiO_2 + 2OH^- \rightarrow SiO_3^{2-} + H_2O$$

SiO_3^{2-} is the silicate ion.

An example of a reaction is:

$$SiO_2 + 2NaOH \rightarrow Na_2SiO_3 + H_2O$$

Na_2SiO_3 is sodium silicate.

Other reactions of SiO_2 directly with basic oxides may be asked about in the exam. The salt formed is a metal silicate and contains the SiO_3^{2-} anion. If hydrogen is present on the left, water is included on the right. These questions are common.

Exam tip

Aluminium foil does not react readily because it forms a protective layer of aluminium oxide on its surface, which prevents further reaction. Aluminium is a reasonably reactive element, but this protective oxide layer often hides its reactivity.

An amphoteric oxide reacts with an acid and with a base, forming a salt.

Exam tip

The aluminate ion can be written in various ways — $Al(OH)_4^-$ or AlO_2^- or $Al(OH)_6^{3-}$ — but $Al(OH)_4^-$ is the most common on AQA papers and mark schemes.

Exam tip

The three oxides so far have all been ionic solids. From here on the oxides become giant covalent and then simple covalent. This change in bonding and structure alters the physical and chemical properties of the oxides.

An acidic oxide reacts with a base, forming a salt.

Worked example

Write an equation for the reaction of sodium oxide with silicon dioxide.

Answer

The salt formed is sodium silicate, Na_2SiO_3. The equation is:

$$Na_2O + SiO_2 \rightarrow Na_2SiO_3$$

Phosphorus

Phosphorus ignites spontaneously in air, burning with a white flame and producing a white solid, P_4O_{10} (phosphorus(v) oxide):

$$P_4 + 5O_2 \rightarrow P_4O_{10}$$

Some phosphorus(III) oxide, P_4O_6, may be produced if the supply of oxygen is limited.

P_4O_{10} is a **molecular covalent acidic oxide**. Its melting point is 300°C owing to the weak intermolecular forces of attraction, which require little energy to break them.

P_4O_{10} reacts with water, producing phosphoric(v) acid, H_3PO_4:

$$P_4O_{10} + 6H_2O \rightarrow 4H_3PO_4$$

The pH of the resulting solution is between 0 and 2. The structure of phosphoric(v) acid is shown in Figure 1.

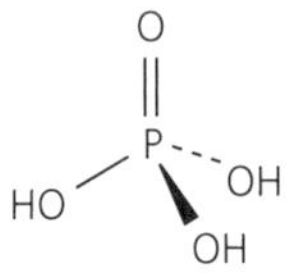

Figure 1 Phosphoric(v) acid

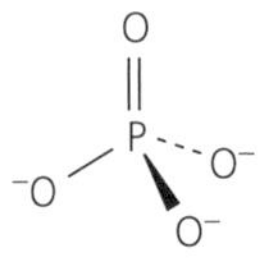

Figure 2 Phosphate(v) ion, PO_4^{3-}

The phosphate(v) ion is formed from the removal of three hydrogen ions. Its structure is shown in Figure 2.

P_4O_{10} may react directly with a basic oxide such as sodium oxide or magnesium oxide. It forms phosphate(v) salt, which contains the anion, PO_4^{3-}.

Exam tip

Always form the salt using the metal cation and the phosphate(v) anion and then balance the equation. These equations can seem complicated, but just remember that the salt formed will contain the metal ion and the phosphate(v) ion. If there is hydrogen on the left, water can be included on the right. Balance the phosphorus atoms first and the rest should be relatively straightforward.

Exam tip

P_4O_{10} is the molecular formula of phosphorus(v) oxide; its empirical formula is sometimes used, which is P_2O_5, but the structure of a molecule of P_4O_{10} contains four P atoms and ten O atoms.

Exam tip

Phosphoric(v) acid also forms the dihydrogenphosphate(v) ion, $H_2PO_4^-$, and the hydrogenphosphate(v) ion, HPO_4^{2-}, with the loss of one or two hydrogen ions respectively. All the ions have the same basic tetrahedral shape.

Worked example

Write an equation for the reaction of P_4O_{10} with NaOH and also with MgO.

Answer

P_4O_{10} reacts with NaOH to form sodium phosphate(v), Na_3PO_4. As hydrogen is present on the left, water is also formed in this neutralisation reaction:

$$12NaOH + P_4O_{10} \rightarrow 4Na_3PO_4 + 6H_2O$$

For MgO, the salt formed is magnesium phosphate(v), which is $Mg_3(PO_4)_2$. No water is required, but always balance the phosphorus atoms first and then the rest will fall into place:

$$6MgO + P_4O_{10} \rightarrow 2Mg_3(PO_4)_2$$

Sulfur

Sulfur burns with a blue flame when heated in air (bluer flame in pure oxygen), releases heat and produces misty fumes of a pungent gas, sulfur dioxide, SO_2:

$$S + O_2 \rightarrow SO_2$$

Sulfur dioxide is also known as sulfur(IV) oxide. It is a **molecular covalent acidic oxide** with a melting point of $-73°C$. The weak intermolecular forces of attraction do not require a lot of energy to break.

Sulfur dioxide reacts with water, producing sulfurous acid, H_2SO_3:

$$SO_2 + H_2O \rightarrow H_2SO_3$$

Sulfurous acid is also known as sulfuric(IV) acid. It is a weak acid and the solution formed has a pH of around 3–4.

SO_2 and H_2SO_3 react with basic oxides and amphoteric oxides, forming salts that contain the sulfite ion, SO_3^{2-}. This is also called the sulfate(IV) ion.

The structures of H_2SO_3 and the sulfate(IV) ion are shown in Figures 3 and 4.

Sulfur dioxide can be converted catalytically to sulfur trioxide, SO_3. Sulfur trioxide is also known as sulfur(VI) oxide. It is a **molecular covalent acidic oxide**. Its melting point is $17°C$. Again, the weak intermolecular forces of attraction require little energy to break.

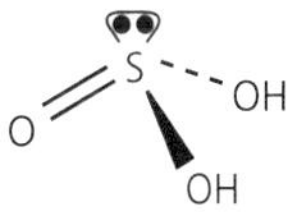

Sulfur trioxide reacts vigorously with water, producing sulfuric acid, H_2SO_4:

$$SO_3 + H_2O \rightarrow H_2SO_4$$

Sulfuric acid is also known as sulfuric(VI) acid. The pH of the resulting solution should be in the range 0–2 because sulfuric acid is a strong acid.

Sometimes a question will ask for an equation that shows the ions formed when sulfur dioxide or sulfur trioxide dissolves in water. $2H^+$ ions and either the sulfite or sulfate ions are produced, respectively:

$$SO_2 + H_2O \rightarrow 2H^+ + SO_3^{2-}$$

$$SO_3 + H_2O \rightarrow 2H^+ + SO_4^{2-}$$

Exam tip

Question on this are common and may use oxides from different periods, for example CaO instead of MgO or K_2O in place of Na_2O. Follow the rules and the equations are straightforward if you remember the anion formed from the acidic oxides.

Knowledge check 2

Write an equation for the reaction of calcium oxide with phosphorus(v) oxide.

Figure 3 Sulfuric(IV) acid, H_2SO_3

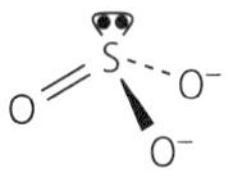

Figure 4 Sulfate(IV) ion, SO_3^{2-}

SO_3 and H_2SO_4 react with basic oxides and amphoteric oxides, forming salts that contain the sulfate ion, SO_4^{2-}. This is also called the sulfate(VI) ion.

The structures of H_2SO_4 and the sulfate(VI) ion are shown in Figures 5 and 6.

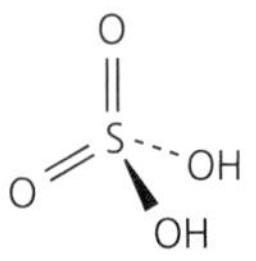

Figure 5 Sulfuric(VI) acid, H_2SO_4

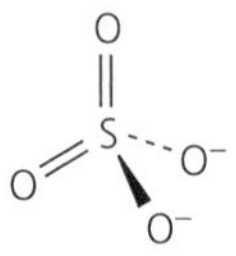

Figure 6 Sulfate(VI) ion, SO_4^{2-}

Exam tip

Equations for the reactions of SO_2 and SO_3 directly with basic oxides can be asked about in the exam. Remember that SO_2 forms salts called sulfites, which contain the SO_3^{2-} anion. SO_3 forms salts called sulfates, which contain the SO_4^{2-} anion. The same rules apply to the equations.

Exam tip

Check the oxidation number carefully, particularly for salts and acids of *p*-block elements: sulfate(VI) is sulfate (SO_4^{2-}) whereas sulfate(IV) is sulfite (SO_3^{2-}). Sulfuric(VI) acid is H_2SO_4 whereas sulfuric(IV) acid is H_2SO_3. These are the IUPAC names and should be given when requested.

Worked example

Write an equation for the reaction of sodium hydroxide with sulfur dioxide.

Answer

The salt formed is sodium sulfite, Na_2SO_3. The equation is:

$$2NaOH + SO_2 \rightarrow Na_2SO_3 + H_2O$$

Trends in melting points of the period 3 oxides

Table 1 shows the melting points of the period 3 oxides.

Table 1 Melting points of the period 3 oxides

Oxide	Melting point/°C
Na_2O	1438
MgO	2852
Al_2O_3	2072
SiO_2	1610
P_4O_{10}	300
SO_3	17

Na_2O, MgO and Al_2O_3 are ionic oxides and a lot of energy is required to break the strong forces of attraction between the ions. SiO_2 is a macromolecular (giant covalent) oxide and again a lot of energy is required to break the many strong covalent bonds in the structure. P_4O_{10} and SO_3 are molecular covalent oxides and little energy is required to break the weak intermolecular forces of attraction. P_4O_{10} has a higher M_r than SO_3, so the forces of attraction between the molecules are stronger.

Knowledge check 3

Write an equation for the reaction of magnesium oxide with sulfur(VI) oxide. Give the IUPAC name of the salt formed.

Knowledge check 4

Name one oxide of a period 3 element that reacts with water to form a strongly alkaline solution.

Summary

- Oxides of metals are either basic or amphoteric.
- Oxides of non-metals are usually acidic.
- Basic oxides (such as Na_2O and MgO) react with acids to form salts; if they react with water they form alkaline solutions.
- Amphoteric oxides (Al_2O_3) react with acids and bases; they form salts with both.
- Acidic oxides react with bases to form salts; if they react with water they form acidic solutions.
- The melting points of the oxides of the elements of period 3 increase from Na_2O to MgO and then start to decrease again.

Transition metals

The series of elements from scandium to zinc is often referred to as the **transition metals**. This is not strictly true, though they are *d*-block elements. A transition metal forms at least one ion with an incomplete *d* subshell. Sc^{3+} and Zn^{2+} are the only ions formed by scandium and zinc, and Sc^{3+} has an empty 3*d* subshell ($1s^2\, 2s^2\, 2p^6\, 3s^2\, 3p^6$) and Zn^{2+} has a complete 3*d* subshell ($1s^2\, 2s^2\, 2p^6\, 3s^2\, 3p^6\, 3d^{10}$). This means that we generally consider Ti to Cu as being the first transition series, while Sc and Zn are *d*-block elements. Chromium and copper atoms show unusual electronic configurations. A chromium atom is $1s^2\, 2s^2\, 2p^6\, 3s^2\, 3p^6\, 3d^5\, 4s^1$ and a copper atom is $1s^2\, 2s^2\, 2p^6\, 3s^2\, 3p^6\, 3d^{10}\, 4s^1$.

General properties of transition metals

Transition metals show the following general properties:

- They form complexes.
- They form coloured ions.
- They have variable oxidation states.
- They show catalytic activity.

Complex formation and shapes of complexes

A complex is a central metal atom or ion surrounded by **ligands**. A ligand is a molecule or ion that forms a coordinate bond with a transition metal atom or ion by donating a pair of electrons.

> **Exam tip**
>
> Transition metal atoms and ions have empty orbitals in the outer energy level, which allow molecules and ions with lone pairs of electrons to form coordinate bonds with these atoms and ions.

A ligand may be **monodentate**; examples are H_2O, NH_3 and Cl^-. These ligands can only form one coordinate bond to the central metal atom or ion.

A ligand may be **bidentate**; examples are 1,2-diaminoethane ($H_2NCH_2CH_2NH_2$) and the ethanedioate ion, $C_2O_4^{2-}$ (Figure 7).

A ligand may be **multidentate**; an example is $EDTA^{4-}$ (Figure 8).

A **transition metal** is an element that forms at least one ion with an incomplete *d* subshell.

Knowledge check 5

What is the electronic configuration of:

a an iron atom
b an Fe^{2+} ion
c an Fe^{3+} ion?

A **ligand** is a molecule or ion that forms a coordinate bond with a transition metal atom or ion.

A **monodentate** ligand forms one coordinate bond with a transition metal atom or ion.

A **bidentate** ligand forms two coordinate bonds with a transition metal atom or ion.

A **multidentate** ligand forms many coordinate bonds with a transition metal atom or ion.

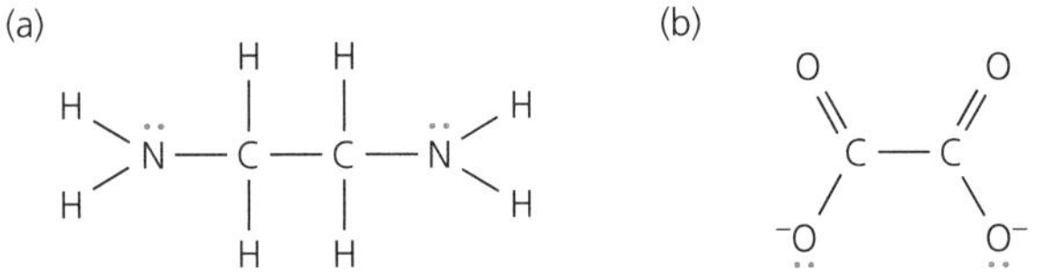

(a)

(b)

Figure 7 (a) 1,2-diaminoethane and (b) the ethanedioate ion showing the lone pairs which form coordinate bonds in a complex

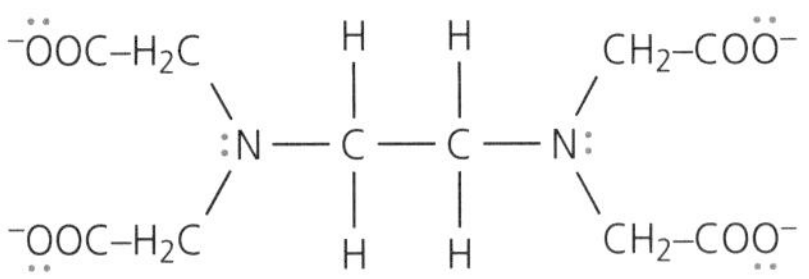

The six lone pairs are shown on the diagram. These form coordinate bonds with the central metal ion. This makes EDTA multidentate and hexadentate.

Figure 8 EDTA^{4-}

The first two ligands have two lone pairs of electrons, which form coordinate bonds to the central metal atom or ion. EDTA^{4-} has six lone pairs of electrons, so one EDTA^{4-} ion can form six coordinate bonds to the central metal atom or ion. It is sometimes referred to as hexadentate.

Examples of complexes and their features

When ligands bind to a transition metal atom or ion a complex is formed. There are various features of the complex that you need to be able to present or describe:

- the formula of the complex
- the oxidation state of the transition metal in the complex
- the coordinate number of the complex
- the shape of the complex
- a diagram of the shape

Table 2 gives some common complexes.

Exam tip

Even though water has two lone pairs of electrons, it is monodentate because the lone pairs are too close together (being on the same atom). As you will see from shapes of complexes, the coordinate bonds form in specific orientations around the atom or ion.

Table 2 Properties of common complexes

Formula	Oxidation state of the transition metal	Coordination number of the complex	Shape of the complex	Diagram of the complex
$[Fe(H_2O)_6]^{3+}$	+3	6	Octahedral	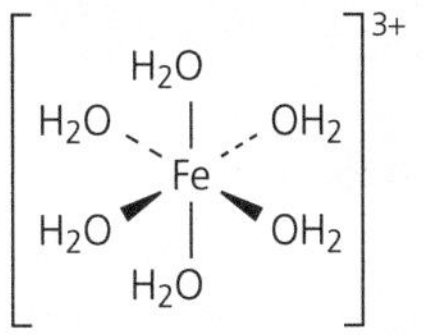
$[Co(NH_3)_6]^{2+}$	+2	6	Octahedral	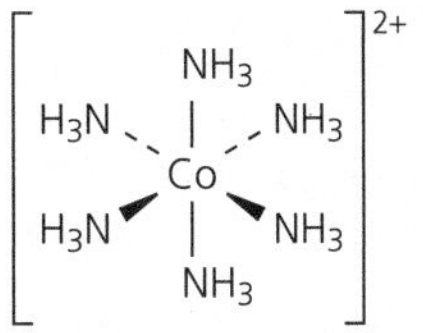
$[Cu(NH_3)_4(H_2O)_2]^{2+}$	+2	6	Octahedral	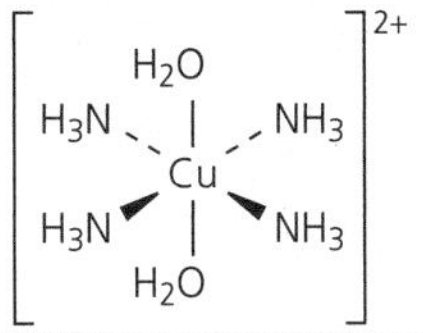

Formula	Oxidation state of the transition metal	Coordination number of the complex	Shape of the complex	Diagram of the complex
$[CoEDTA]^{2-}$	+2	6	Octahedral	
$[Co(H_2NCH_2CH_2NH_2)_3]^{2+}$	+2	6	Octahedral	
$[CuCl_4]^{2-}$	+2	4	Tetrahedral	
$[FeCl_4]^-$	+3	4	Tetrahedral	
$[Pt(NH_3)_2Cl_2]$	+2	4	Square planar	
$[Ag(NH_3)_2]^+$	+1	2	Linear	

From Table 2:

1 Many transition metal ions form complexes with water. These are of the form $[M(H_2O)_6]^{n+}$, where M represents the transition metal and $n+$ is the charge on the complex ion. These are called hexaaqua cations.

Water is a small ligand, so six ligands can form coordinate bonds around the central metal ion.

For larger ligands such as Cl^- in $[CuCl_4]^{2-}$ and $[FeCl_4]^-$ only four can form coordinate bonds as more would repel each other.

2 The oxidation state of the transition metal in the complex together with the charge on any ligands gives the overall charge on the complex. Where the ligands are neutral, such as in $[Co(NH_3)_6]^{2+}$, the charge on the ion is the same as the oxidation state of the transition metal.

For $[FeCl_4]^-$, each chloro ligand is Cl^-, so to obtain an overall charge of $-$, the iron must be in the +3 oxidation state.

Similarly, in $[Pt(NH_3)_2Cl_2]$, the NH_3 has no charge, but the Cl^- means that no overall charge on the complex gives the Pt a +2 oxidation state.

In $[CoEDTA]^{2-}$, $EDTA^{4-}$ means that the cobalt has an oxidation state of +2.

3 The coordination number of a complex is the number of coordinate bonds to the central metal atom or ion. It is not simply the total number of ligands unless all the ligands are monodentate.

For $[CoEDTA]^{2-}$, $EDTA^{4-}$ forms six coordinate bonds to the Co^{2+} ion, so the coordination number is 6 even though there is only one ligand ion.

For $[Co(H_2NCH_2CH_2NH_2)_3]^{2+}$ three bidentate ligands are coordinately bonded to the central Co^{2+} ion. Again, the coordination number is 6.

4 There are four main shapes of complexes — linear, octahedral, square planar and tetrahedral.

Complexes with a coordination number of 6 are octahedral.

Complexes with a coordination number of 4 are usually tetrahedral, but Pt complexes are square planar.

Complexes with a coordination number of 2 are linear.

5 The shapes of complexes should be drawn similarly to the shapes of molecules studied at AS.

Any bonds coming out of the plane of the paper should be drawn as wedges.

Any going into the plane of the paper should be drawn as dashed lines.

This is most common with octahedral and tetrahedral complexes.

6 Some complexes may contain more than one type of ligand. $[Cu(NH_3)_4(H_2O)_2]^{2+}$ is an example. This is formed when a ligand replacement reaction is not complete. These are examined in the next section.

Exam tip

It is unlikely that you would be expected to draw the full structure of the complexes with multidentate ligands. Focus on sketching the shapes of the simpler ones with monodentate ligands.

Exam tip

The shapes of complexes can be remembered using LOST: L is linear, O is octahedral, S is square planar and T is tetrahedral.

Knowledge check 6

State the shape and coordination number of the following complexes:

a $[Co(H_2O)_6]^{2+}$
b $[Ag(NH_3)_2]^+$
c $[Pt(NH_3)_2Cl_2]$

Ligand substitution reactions

When a complex reacts with a solution containing a new ligand, a ligand substitution reaction may occur. The feasibility of this ligand substitution reaction depends on the enthalpy change of the reaction and the entropy change.

Ligand substitution with no change in coordination number

In the reaction below, ammonia is replaced by 1,2-diaminoethane as a ligand in a cobalt complex. The coordination number remains the same.

$$[Co(NH_3)_6]^{2+}(aq) + 3H_2NCH_2CH_2NH_2(aq) \rightarrow$$
$$[Co(H_2NCH_2CH_2NH_2)_3]^{2+}(aq) + 6NH_3(aq)$$

On the left-hand side there are four different particles (one complex and three free ligands) and on the right-hand side there are seven different particles (one complex and six free ligands). This represents an increase in the number of particles in solution, so there is an increase in entropy. Also, the same number of coordinate bonds are being broken and formed, so the enthalpy change will be approximately zero. Therefore, overall, ΔG will be negative and so the reaction is feasible.

Chelate effect

A **chelate** is the complex formed between a transition metal atom or ion and a multidentate ligand. Chelates are inherently stable. The chelate effect is caused by the increase in entropy and results in a feasible reaction when a multidentate ligand replaces a monodentate ligand.

One small ligand may be replaced by another small ligand, for example NH_3 replacing H_2O in the following ligand replacement reaction:

$$[Co(H_2O)_6]^{2+}(aq) + 6NH_3(aq) \rightarrow [Co(NH_3)_6]^{2+}(aq) + 6H_2O(l)$$

Here the same number of coordinate bonds are broken and made, and there is no increase or decrease in the number of particles in solution, yet the reaction occurs. The $[Co(NH_3)_6]^{2+}$ complex is more stable than the $[Co(H_2O)_6]^{2+}$ complex. NH_3 is a stronger ligand than H_2O and forms stronger coordinate bonds, so ΔH is negative.

Ligand substitution with a change in coordination number

In the following reaction, again one small ligand is replaced with another but there is a change in coordination number:

$$[Co(H_2O)_6]^{2+}(aq) + 4Cl^-(aq) \rightarrow [CoCl_4]^{2-}(aq) + 6H_2O(l)$$

The coordination number changes from 6 to 4. This means that six coordinate bonds are broken and four are formed but, more importantly, there are five particles in solution on the left-hand side and seven in solution on the right-hand side, so there is an increase in entropy.

Haem

Haem is an iron(II) complex with a multidentate ligand. It is found at the centre of haemoglobin, the protein that transports oxygen in blood. The structure of haem is shown in Figure 9.

Chelate A complex formed between a transition metal atom or ion and a multidentate ligand.

Knowledge check 7

For the reaction $[Fe(H_2O)_6]^{2+} + 4Cl^- \rightarrow [FeCl_4]^{2-} + 6H_2O$, explain why there is an increase in entropy.

Four nitrogen atoms form coordinate bonds to the central Fe^{2+} ion. The ring structure with the four nitrogen atoms is called a porphyrin ring. The complex of these four N atoms with the Fe^{2+} is square planar. Haemoglobin has this structure and four proteins combined. A fifth coordinate bond comes from an amino acid residue in one of the protein chains. The sixth coordinate bond position is for O_2 molecules, which have a lone pair of electrons and can form a coordinate bond with the complex. The complex formed when O_2 binds to haemoglobin is called oxyhaemoglobin.

Carbon monoxide also has a lone pair of electrons on the oxygen atom and so can bind to haemoglobin. It is a stronger ligand and binds more strongly than oxygen so less and less oxygen is carried in the blood, often resulting in death. The complex formed between carbon monoxide and haemoglobin is called carboxyhaemoglobin.

Stereoisomerism in complexes

Some complexes show E–Z isomerism and others show optical isomerism.

E–Z isomerism

Complexes of the formula ML_2A_2 (where M represents the transition metal atom or ion and L and A represent different monodentate ligands) form E and Z complexes depending on the 3D spatial arrangement of the ligands. The main example of this type of isomerism is shown by $[Pt(NH_3)_2Cl_2]$. The structure of the two isomers is shown in Figure 10. Often the Z form is called *cis* and the E form is called *trans*. The *cis* form is used as an anticancer drug and is called cisplatin (Figure 10). Transplatin has no anticancer activity.

$$
\begin{array}{ccc}
& \text{Cl} & \\
& | & \\
\text{H}_3\text{N} - &\text{Pt} &- \text{Cl} \\
& | & \\
& \text{NH}_3 &
\end{array}
\qquad
\begin{array}{ccc}
& \text{NH}_3 & \\
& | & \\
\text{Cl} - &\text{Pt} &- \text{Cl} \\
& | & \\
& \text{NH}_3 &
\end{array}
$$

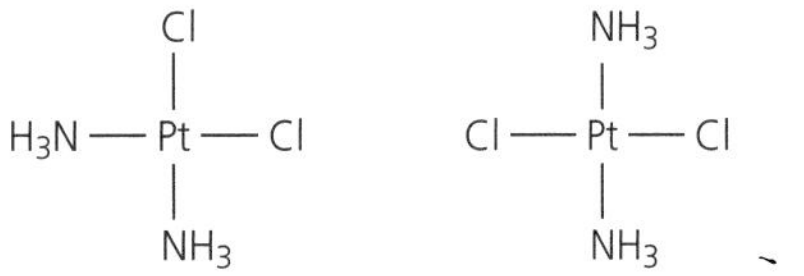

Z isomer (*cis* form) E isomer (*trans* form)

Figure 10 E and Z isomers of $[Pt(NH_3)_2 Cl_2]$

In the Z form the two higher-priority chloro ligands are beside each other, whereas in the E form they are opposite each other.

Octahedral complexes with the formula ML_4A_2 (where M represents the transition metal atom or ion and L and A represent different monodentate ligands), such as $[Co(NH_3)_4Cl_2]^+$, also form E and Z isomers. In this complex the cobalt has an oxidation state of +3. There are two distinct forms of the octahedral complex where the higher-priority chloro ligands are beside each other (Z form) or where they are opposite each other (E form). Figure 11 shows examples of the complexes.

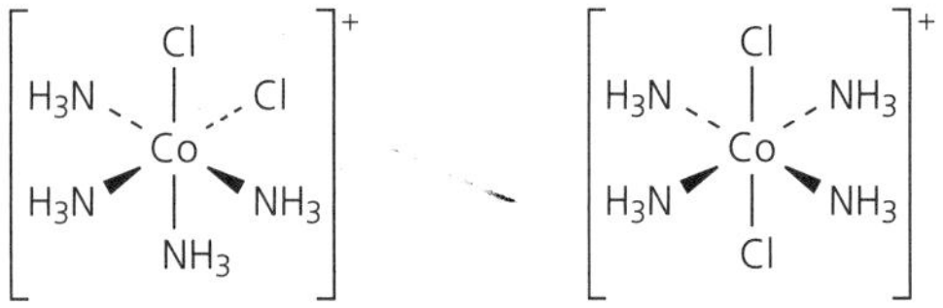

Z isomer E isomer

Figure 11 E and Z isomers of $[Co(NH_3)_4Cl_2]^+$

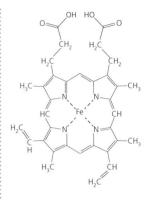

Figure 9 The structure of haem

Knowledge check 8

State the shape of the complex formed between the porphyrin ring and Fe^{2+} in haem.

Optical isomerism

Octahedral complexes containing a bidentate ligand can form optical isomers. These two isomers are non-superimposable, based on the 3D spatial arrangement of the ligands around the transitional metal ion. An example is $[Co(H_2NCH_2CH_2NH_2)_3]^{3+}$. Figure 12 shows the two optical isomers where $H_2NCH_2CH_2NH_2$ is represented by $H_2N–NH_2$. The green line represents $–CH_2–CH_2–$.

Knowledge check 9

State the medical use of the *Z* isomer of $[Pt(NH_3)_2Cl_2]$.

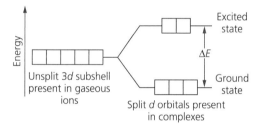

Figure 12 Two optical isomers of $[Co(H_2NCH_2CH_2NH_2)_3]^{3+}$

Formation of coloured ions

Transition metal ions in solution form coloured complexes. This is due to the five $3d$ orbitals splitting into two distinct sets, which are separated by an energy difference referred to as ΔE (Figure 13).

Figure 13 Splitting of $3d$ orbitals

The energy difference allows the complex to absorb some light from the visible region of the electromagnetic spectrum. The light absorbed excites the electrons from the lower orbitals to the higher ones. The observed or transmitted colour is the complementary colour to the colour(s) that are absorbed.

Table 3 shows the wavelength of light in the visible region of the spectrum and the colours of light absorbed and observed.

Table 3 Absorption of colours

Wavelength/nm	Colour absorbed	Colour observed
400–430	Violet	Yellow–green
430–460	Blue–violet	Yellow
460–490	Blue	Orange
490–510	Blue–green	Red
510–530	Green	Purple
530–560	Yellow–green	Violet
560–590	Yellow	Blue–violet
590–610	Orange	Blue
610–700	Red	Blue–green

The colour(s) of visible light absorbed depend(s) on:

- the metal in the complex
- the oxidation state of the metal in the complex
- the coordination number of the complex
- the shape of the complex

ΔE, frequency and wavelength

Figure 14 shows the links between the energy of the light absorbed (ΔE), the frequency of the light (ν) and the wavelength of the light (λ).

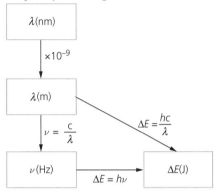

Figure 14 The links between ΔE, frequency and wavelength

The key to calculations involving ΔE, frequency and wavelength is understanding the units:

- ΔE is measured in joules (J).
- Frequency (ν) is measured in hertz (Hz), which is the same as s^{-1}.
- Wavelength (λ) is often measured in nanometres (nm). 1 nm is 10^{-9} m.
- c is the speed of light and it is measured in $m\,s^{-1}$. It is usually quoted as $3.00 \times 10^8\,m\,s^{-1}$.
- h is Planck's constant and it has a value of 6.63×10^{-34} J s.

Worked example

Calculate the value of ΔE for a wavelength of 650 nm.

$h = 6.63 \times 10^{-34}$ J s and $c = 3.00 \times 10^8\,m\,s^{-1}$

Answer

First, convert the wavelength from nm to m by multiplying by 10^{-9}:

$$\lambda = 650 \times 10^{-9}\,m = 6.5 \times 10^{-7}\,m$$

Then use $\Delta E = hc/\lambda$:

$$\frac{6.63 \times 10^{-34} \times 3.00 \times 10^8}{6.50 \times 10^{-7}} = 3.06 \times 10^{-19}\,J$$

These calculations may be given in a different format, or you may have to calculate the wavelength from the ΔE value.

Spectroscopy and colorimetry

The concentration of a coloured species in solution can be determined using spectroscopy. A spectrophotometer allows the measurement of absorbance at selected wavelengths of light from the visible and/or ultraviolet regions of the electromagnetic spectrum. Spectroscopy can detect absorbance for substances that appear as colourless solutions, as the solute may absorb in the ultraviolet region. Once an absorbance is detected a calibration curve can be set up that plots the absorbance against the concentration of the substance in the solution. This curve allows absorbances of the other solutions to be converted to concentrations.

A typical calibration curve is shown in Figure 15.

For coloured substances a colorimeter may be used that measures absorbance of coloured light. A colorimeter has a source light of a selected wavelength (or colour) and passes it though a solution. The amount of light absorbed is proportional to the concentration of the coloured species in the solution. The selection of the wavelength or colour of light is important. For example, a blue solution does not absorb blue light, so the filter used should be red, to select red light. The solution sample is placed in a cuvette, which is a 1 cm × 1 cm vial (Figure 16).

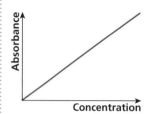

Figure 15 A calibration curve

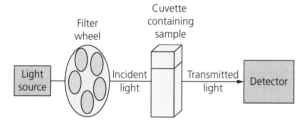

Figure 16 A colorimeter

Variable oxidation states

Transition metals show variable oxidation states. Table 4 shows the oxidation states of the first transition series from Ti to Cu.

Table 4 Oxidation states of the first transition series

Element	Ti	V	Cr	Mn	Fe	Co	Ni	Cu
				+7				
			+6	+6	+6			
		+5	+5	+5	+5			
Oxidation states	+4	+4	+4	+4	+4	+4	+4	
	+3	+3	+3	+3	+3	+3	+3	+3
	+2	+2	+2	+2	+2	+2	+2	+2
			+1					+1

The higher oxidation states are found in covalent compounds and compound ions. Simple ions with charges higher than 3+ are rare.

Vanadium

Vanadium can exist in the +2, +3, +4 and +5 oxidation states in various compounds and ions. Table 5 shows some of the compound ions with these oxidation states.

Content Guidance

Table 5 Vanadium compounds

Oxidation state	Name of molecular ion	Formula of ion	Colour in aqueous solution
+5	Dioxovanadium(v) ion	VO_2^+	Yellow
	Vanadate(v) ion	VO_3^-	
+4	Oxovanadium(iv) ion	VO^{2+}	Blue
+3	Vanadium(iii) ion	V^{3+}	Green
+2	Vanadium(ii) ion	V^{2+}	Violet

Exam tip

The colours of the ions in solution are also important, so you should devise a way to remember them, such as **Y**ou **B**etter **G**et **V**anadium. Make up your own, as long as you can remember it!

It is important to be able to recognise and name compounds of these ions. For example, the IUPAC name for $VOSO_4$ is oxovanadium(iv) sulfate(vi). NH_4VO_3 is ammonium vanadate(v).

Reduction of vanadium from +5 to +2

Vanadium in the +5 oxidation state can be reduced to the +2 oxidation state using zinc in acidic solution, usually in the presence of dilute hydrochloric acid.

The equations for the reduction are given below, with their standard electrode potentials.

+5 to +4: $VO_2^+ + 2H^+ + e^- \rightarrow VO^{2+} + H_2O$ $E^\ominus = +1.00\,V$
      yellow solution   blue solution

+4 to +3: $VO^{2+} + 2H^+ + e^- \rightarrow V^{3+} + H_2O$ $E^\ominus = +0.32\,V$
      blue solution   green solution

+3 to +2: $V^{3+} + e^- \rightarrow V^{2+}$ $E^\ominus = -0.26\,V$
      green solution   violet solution

The standard electrode potential for the reduction of zinc ions is:

$Zn^{2+} + 2e^- \rightarrow Zn$ $E^\ominus = -0.76\,V$

Knowledge check 11

Write the formula for dioxovanadium(v) chloride.

As the three standard electrode potentials for vanadium to reduce from +5 to +2 are greater than $E^\ominus$ (Zn^{2+}/Zn), zinc can reduce vanadium from +5 to +2. The acidic conditions provide the H^+ ions. Calculating the EMF for each reaction would give +1.76 V for the +5 to +4 reduction, +1.08 V for the +4 to +3 reduction and +0.50 V for the +3 to +2 reduction. All these reactions are feasible, so zinc will reduce vanadium from +5 to +2.

Exam tip

As the reduction occurs the initial yellow solution often changes to green before becoming blue. This green colour is a mixture of the yellow and the blue and should not be confused with the +3 oxidation state of vanadium.

Worked example

The reduction of vanadium occurs in a series of steps:

$VO_2^+ + 2H^+ + e^- \rightarrow VO^{2+} + H_2O$ $E^\ominus = +1.00\,V$

$VO^{2+} + 2H^+ + e^- \rightarrow V^{3+} + H_2O$ $E^\ominus = +0.32\,V$

$V^{3+} + e^- \rightarrow V^{2+}$ $E^\ominus = -0.26\,V$

Considering the standard electrode potentials given below:

$Fe^{2+}(aq) + 2e^- \rightarrow Fe(s)$ $E^\ominus = -0.44\,V$

$Fe^{3+}(aq) + e^- \rightarrow Fe^{2+}(aq)$ $E^\ominus = +0.77\,V$

$$Sn^{2+}(aq) + 2e^- \rightarrow Sn(s) \qquad E^{\ominus} = -0.14\,V$$

$$Ni^{2+}(aq) + 2e^- \rightarrow Ni(s) \qquad E^{\ominus} = -0.25\,V$$

Which one of the following will reduce vanadium from +5 to +4, but no further?

A   iron   C   tin

B   iron(II) ions   D   nickel

Answer

The answer is B.

The +5 to +4 reduction of vanadium has a standard electrode potential of +1.00 V. The reduction of iron(III) ions to iron(II) ions is less than this but greater than the other two electrode potentials of vanadium, so the only reaction that will occur is the reduction from +5 to +4. The EMF is +0.23 V, whereas the reduction from +4 to +3 would give an EMF of −0.45 V and +3 to +2 would be −1.03 V. Check that you can calculate these values.

Iron would reduce vanadium from +5 to +2 (EMF values are +1.44 V, +0.76 V and +0.18 V); tin would reduce vanadium from +5 to +3 (EMF values are +1.14 V, +0.46 V and −0.12 V); nickel would reduce vanadium from +5 to +3 as well (EMF values are +1.25 V, + 0.57 V and −0.01 V). The EMF values are for the possible reductions from +5 to +4, +4 to +3 and +3 to +2 respectively. Again, check that you would have got these values.

Variations in electrode potentials

Changes in pH and ligands can affect the value for the electrode potential for the reduction of a transition metal.

The reduction of silver(I) ions is:

$$Ag^+(aq) + e^- \rightarrow Ag(s) \qquad\qquad\qquad E^{\ominus} = +0.80\,V$$

However, in Tollens' reagent $[Ag(NH_3)_2]^+$ ions are present. The reduction to silver is given by:

$$[Ag(NH_3)_2]^+(aq) + e^- \rightarrow Ag(s) + 2NH_3(aq) \qquad E^{\ominus} = +0.37\,V$$

Silver(I) ions are complexed with ammonia in Tollens' reagent to control the reaction. The complex is a milder oxidising agent than aqueous silver(I) ions. Silver(I) ions would cause a faster reaction and silver would appear as a solid precipitate in the solution, making it cloudy. The slower reaction with the complex allows the formation of a silver mirror.

Changes in pH can also affect the electrode potentials. Acidic conditions allow manganate(VII) ions, MnO_4^-, to be reduced to manganese(II) ions with an electrode potential of +1.51 V. Alkaline conditions (usually achieved using sodium carbonate solution) will cause the manganate(VII) ions, MnO_4^-, to be reduced to manganate(VI) ions, MnO_4^{2-}, with an electrode potential of +0.60 V. The equations are given below:

$$MnO_4^- + 8H^+ + 5e^- \rightarrow Mn^{2+} + 4H_2O \quad E^{\ominus} = +1.51\,V$$

$$MnO_4^- + e^- \rightarrow MnO_4^{2-} \qquad\qquad\qquad E^{\ominus} = +0.60\,V$$

Knowledge check 12

Write a redox equation for the reduction of vanadium(III) ions to vanadium(II) ions using zinc.

Manganate(VII) ions in acidic solution form a very powerful oxidising agent, which can completely oxidise organic molecules, breaking carbon–carbon bonds. An alkaline solution of manganate(VII) is a milder oxidising agent, which will oxidise alkenes to diols. Manganate(VII) ions are purple in solution, manganate(VI) ions are dark green and manganese(II) ions are colourless. Alkaline potassium manganate(VII) can be used for a test for unsaturation. On prolonged reaction manganate(VI) ions are further reduced to a dark brown precipitate of manganese(IV) oxide, MnO_2.

Redox titrations

A solution containing manganate(VII) ions, MnO_4^-, will react with a reducing agent such as iron(II) ions, Fe^{2+}, or ethanedioate ions, $C_2O_4^{2-}$. The purple solution containing manganate(VII) ions is added to a conical flask containing the reducing agent. As the solution is added it changes colour from purple to colourless as the manganate(VII) ions are reduced to manganese(II) ions, Mn^{2+}:

$$MnO_4^- + 8H^+ + 5e^- \rightarrow Mn^{2+} + 4H_2O$$

purple solution   colourless solution

The reducing agent is oxidised. Iron(II) ions are oxidised to iron(III) ions and ethanedioate ions are oxidised to carbon dioxide:

$$Fe^{2+} \rightarrow Fe^{3+} + e^-$$

$$C_2O_4^{2-} \rightarrow 2CO_2 + 2e^-$$

The overall redox equations are obtained by balancing the electrons:

$$MnO_4^- + 8H^+ + 5Fe^{2+} \rightarrow Mn^{2+} + 5Fe^{3+} + 4H_2O$$

$$2MnO_4^- + 16H^+ + 5C_2O_4^{2-} \rightarrow 2Mn^{2+} + 10CO_2 + 8H_2O$$

The overall ratios of MnO_4^- to Fe^{2+} and MnO_4^- to $C_2O_4^{2-}$ are 1:5 and 2:5 respectively. It is important to remember these as it will save you having to work out the overall equation if not asked to do it as part of a question.

The examples that follow show the basic types of redox titration calculations.

> **Exam tip**
>
> The ratio of MnO_4^-:Fe^{2+} is 1:5. It is important to remember this ratio and use it correctly in the question. Most mistakes are made by the incorrect use of a ratio. Three significant figures are maintained throughout this calculation.

Worked example 1

A sample of 7.78 g of hydrated iron(II) sulfate, $FeSO_4.xH_2O$, was dissolved in deionised water and transferred to a volumetric flask where the volume was made up to 250 cm^3. A 25.0 cm^3 sample of this solution was titrated with 0.0250 mol dm^{-3} potassium manganate(VII) solution. The titre was 22.40 cm^3. Calculate the value of x in $FeSO_4.xH_2O$.

Answer

$$\text{moles of } MnO_4^- = \frac{22.40 \times 0.0250}{1000} = 5.60 \times 10^{-4} \text{ mol}$$

$$\text{moles of } Fe^{2+} \text{ in } 25.0 \text{ cm}^3 = 5.60 \times 10^{-4} \times 5 = 2.80 \times 10^{-3} \text{ mol}$$

$$\text{moles of } Fe^{2+} \text{ in } 250 \text{ cm}^3 = 2.80 \times 10^{-3} \times 10 = 0.0280 \text{ mol}$$

As 1 mol of $FeSO_4.xH_2O$ contains 1 mol of Fe^{2+}, 0.0280 mol of $FeSO_4.xH_2O$ must have been added.

> **Exam tip**
>
> Try the calculation with 23.95 cm^3 as the average titre and the answer for x should be 6. 33.1 cm^3 gives $x = 2$. Also try the calculation the other way round. See if you can calculate the volume of 0.0250 mol dm^{-3} $KMnO_4$ solution required to react if the value of x is known.

$\rightarrow$

$7.78\,g = 0.028\,mol$ of $FeSO_4.xH_2O$

$$M_r = \frac{7.78}{0.0280} = 277.9$$

M_r of $FeSO_4 = 55.8 + 32.1 + 4(16.0) = 151.9$

M_r of $xH_2O = 277.9 - 151.9 = 126.0$

$$x = \frac{126.0}{18.0} = 7$$

Knowledge check 13

15.0 g of an impure sample of anhydrous iron(II) sulfate, $FeSO_4$, were dissolved in $1\,dm^3$ of deionised water in a volumetric flask. A $25.0\,cm^3$ sample required $17.4\,cm^3$ of $0.0220\,mol\,dm^{-3}$ potassium manganate(VII) solution. Calculate the percentage purity of the sample. Give your answer to three significant figures.

Worked example 2

2.00 g of iron(II) ethanedioate, FeC_2O_4, were dissolved in deionised water and the volume made up to $500\,cm^3$ in a volumetric flask. $25.0\,cm^3$ of this solution were titrated with $0.0145\,mol\,dm^{-3}$ potassium manganate(VII) solution. Calculate the volume of potassium manganate(VII) solution required to react completely. Give your answer to three significant figures.

Answer

Iron(II) ethanedioate contains both Fe^{2+} ions and $C_2O_4^{2-}$ ions. Both ions react with manganate(VII) ions:

$$MnO_4^- + 8H^+ + 5Fe^{2+} \rightarrow Mn^{2+} + 5Fe^{3+} + 4H_2O$$

$$2MnO_4^- + 16H^+ + 5C_2O_4^{2-} \rightarrow 2Mn^{2+} + 10CO_2 + 8H_2O$$

3 mol of MnO_4^- react with 5 mol of FeC_2O_4. This ratio is determined by combining the equations as 5 mol of FeC_2O_4 would produce 5 mol of Fe^{2+} and 5 mol of $C_2O_4^{2-}$. So 3 mol of MnO_4^- in total are required to react.

$$\text{mol of } FeC_2O_4 = \frac{2.00}{143.8} = 0.0139\,mol$$

$$\text{mol of } FeC_2O_4 \text{ in } 25.0\,cm^3 = \frac{0.0139}{20} = 6.95 \times 10^{-4}\,mol$$

$$\text{moles of } MnO_4^- \text{ required to react} = \frac{6.95 \times 10^{-4}}{5} \times 3 = 4.17 \times 10^{-4}\,mol$$

$$\text{volume of } KMnO_4 = \frac{4.17 \times 10^{-4} \times 1000}{0.0145} = 28.8\,cm^3$$

Exam tip

Again the answers for each step are given to three significant figures. However, doing the entire calculation using the answers in your calculator will give the same volume. Try the calculation again using 1.50 g, which gives a volume of $21.6\,cm^3$.

Catalysts

A **catalyst** is a substance that provides an alternative reaction route or pathway of lower activation energy. Catalysts are either homogeneous or heterogeneous. A **homogeneous catalyst** is in the same phase (or state) as the reactants. A **heterogeneous catalyst** is in a different phase from the reactants.

Transition metals and their compounds act as both heterogeneous and homogeneous catalysts.

A **catalyst** is a substance that speeds up a reaction and provides a reaction pathway of lower activation energy. A **homogeneous catalyst** is in the same phase (or state) as the reactants in the reaction.

A **heterogeneous catalyst** is in a different phase (or state) from the reactants in the reaction.

Heterogeneous catalysts

Examples of heterogeneous catalysts are:

- iron in the Haber process
- vanadium(v) oxide in the Contact process
- nickel in the hydrogenation of fats

Heterogeneous catalysts, such as iron and nickel above, function by reactant molecules being adsorbed onto active sites on the surface of the catalyst. The bonds in the reactant molecules are weakened and the molecules are held in a more favourable conformation for reaction. Once the reaction is complete the product molecules are desorbed from the active sites. Many heterogeneous catalysts are expensive metals and so they are often coated onto a honeycomb style support medium to maximise the surface area of the catalyst available and also to minimise cost.

Some heterogeneous catalysts, such as vanadium(v) oxide, function by changing oxidation state. In the Contact process to manufacture sulfuric acid, the conversion of sulfur dioxide into sulfur trioxide is catalysed by vanadium(v) oxide. The vanadium(v) oxide reacts with the sulfur dioxide:

$$SO_2 + V_2O_5 \rightarrow SO_3 + V_2O_4$$

The vanadium(IV) oxide, V_2O_4, then reacts with oxygen to reform vanadium(v) oxide:

$$2V_2O_4 + O_2 \rightarrow 2V_2O_5$$

Combining these equations by multiplying the first one by 2 and adding them together gives the overall equation for the reaction:

$$2SO_2 + O_2 \rightarrow 2SO_3$$

A catalyst can become poisoned by impurities in the reactants. An example of this would be the poisoning of the catalyst in catalytic converters in cars and other vehicles. The catalyst is poisoned by the use of leaded petrol. The lead coats the surface of the catalyst and renders it passive. This has cost implications as the catalyst is made from expensive metals and they would have to be replaced.

Homogeneous catalysts

Aqueous iron(II) or iron(III) ions catalyse the reaction between peroxodisulfate ions, $S_2O_8^{2-}$, and iodide ions, I^-, in solution. The reaction is:

$$S_2O_8^{2-} + 2I^- \rightarrow 2SO_4^{2-} + I_2$$

This reaction is slow because the two negative ions repel each other, but the presence of iron(II) ions or iron(III) ions in solution will speed it up.

The mechanism for the reaction is:

$$2Fe^{2+} + S_2O_8^{2-} \rightarrow 2Fe^{3+} + 2SO_4^{2-}$$

$$2Fe^{3+} + 2I^- \rightarrow 2Fe^{2+} + I_2$$

These reactions can occur in either order, so both Fe^{2+} and Fe^{3+} can catalyse the reaction. Again, the variable oxidation state of iron is important in this catalysis.

Knowledge check 14

Explain why iron is described as a heterogeneous catalyst in the Haber process.

Homogeneous autocatalysis

The reaction between ethanedioate ions and manganate(VII) ions is:

$$2MnO_4^- + 5C_2O_4^{2-} + 16H^+ \rightarrow 2Mn^{2+} + 10CO_2 + 8H_2O$$

The reaction is slow initially as the two negative ions repel each other. However Mn^{2+} ions are formed and these act as a catalyst in the reaction. The mechanism is:

$$4Mn^{2+} + MnO_4^- + 8H^+ \rightarrow 5Mn^{3+} + 4H_2O$$

$$2Mn^{3+} + C_2O_4^{2-} \rightarrow 2Mn^{2+} + 2CO_2$$

The positively charged Mn^{2+} ion is attracted to the MnO_4^- ion. As the reaction proceeds more Mn^{2+} ions are formed, so the reaction rate increases. As MnO_4^- is purple in solution and the colour fades to colourless as the reaction proceeds, the reaction can be monitored using a colorimeter. Figure 17 shows how the concentration of MnO_4^- changes against time. The slope of this line gives the rate of the reaction. The steeper the slope, the faster the reaction.

This type of catalysis is referred to as autocatalysis because the reaction is catalysed by one of the products. Again the variable oxidation state of manganese allows it to act as a catalyst in this reaction.

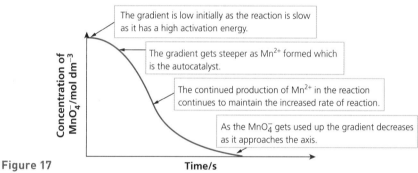

The gradient is low initially as the reaction is slow as it has a high activation energy.

The gradient gets steeper as Mn^{2+} formed which is the autocatalyst.

The continued production of Mn^{2+} in the reaction continues to maintain the increased rate of reaction.

As the MnO_4^- gets used up the gradient decreases as it approaches the axis.

Figure 17

Concentration of MnO_4^-/mol dm^{-3} (y-axis)

Time/s (x-axis)

Summary

- Transition metals form complexes with ligands. A ligand is a molecule or ion that can form a coordinate bond to the transition metal atom or ion
- Ligands can be monodentate, such as H_2O or Cl^-, or multidentate, such as 1,2-diaminoethane, ethanedioate or $EDTA^{4-}$.
- Complexes have the following shapes: linear, octahedral, square planar or tetrahedral.
- Ligand substitution reactions can occur in which the ligands in a complex are replaced. This is often caused by an increase in entropy.
- Many complexes are coloured. This is caused by the splitting of the d subshell.
- The difference in energy between these split levels is represented by $\Delta E = h\nu$, where h is Planck's constant and ν is frequency measured in Hz.

- Stereoisomerism (E–Z and optical) exists in some complexes.
- Vanadium, like many transition metals, shows a variety of oxidation states that have characteristic colours.
- A redox titration using manganate(VII) ions can be used to determine the amount of a reducing agent, such as Fe^{2+} ions or ethanedioate ions, $C_2O_4^{2-}$.
- Catalysts in the same state as the reactants are called homogeneous catalysts. Catalysts in a different state from the reactants are called heterogeneous catalysts.
- Autocatalysis occurs when the product of a reaction acts as a catalyst, for example Mn^{2+} in the reaction between manganate(VII) ions and ethanedioate ions.

Reactions of ions in aqueous solution

Metal ions in solution react with sodium hydroxide solution, ammonia solution and sodium carbonate solution.

Both sodium hydroxide solution and ammonia solution contain hydroxide ions. Many metal hydroxides are insoluble in water, so adding a solution containing hydroxide ions to a solution containing metal ions may produce a precipitate. When excess sodium hydroxide solution or excess ammonia solution are added, some precipitates redissolve as they form a complex.

Sodium carbonate solution contains carbonate ions and many metal carbonates are insoluble in water. Adding sodium carbonate solution to a solution containing transition metal ions may produce a precipitate.

Complexes

Many metal ions form hexaaqua complexes when they dissolve in water. The ones we have to examine are:

- $[Fe(H_2O)_6]^{2+}$
- $[Cu(H_2O)_6]^{2+}$
- $[Fe(H_2O)_6]^{3+}$
- $[Al(H_2O)_6]^{3+}$

A solution containing either $[Fe(H_2O)_6]^{3+}$ or $[Al(H_2O)_6]^{3+}$ ions is acidic. This is due to proton abstraction, where the high charge density of the ion polarises the water ligands and some of these ligands release hydrogen ions into the solution. For example:

$$[Al(H_2O)_6]^{3+} \rightleftharpoons [Al(H_2O)_5OH]^{2+} + H^+$$

When sodium carbonate is added to a solution containing these ions, the carbonate is broken down by the presence of the acid and the precipitate formed is the hydroxide. Each of the ions reacting with each of the three solutions will be considered.

$[Fe(H_2O)_6]^{2+}$

The addition of sodium hydroxide solution or ammonia solution to a solution containing $[Fe(H_2O)_6]^{2+}$ ions produces a green precipitate. The green precipitate is iron(ii) hydroxide. This slowly changes to a brown precipitate.

$$[Fe(H_2O)_6]^{2+} + 2OH^- \rightarrow Fe(OH)_2(H_2O)_4 + 2H_2O$$
$$\text{green ppt}$$

The precipitate does not redissolve on addition of either excess sodium hydroxide solution or excess ammonia solution. The change of colour of the precipitate is due to oxidation of iron(ii) hydroxide to iron(iii) hydroxide by oxygen in the air.

The addition of sodium carbonate solution to a solution containing $[Fe(H_2O)_6]^{2+}$ ions produces a green precipitate of iron(ii) carbonate, $FeCO_3$:

$$[Fe(H_2O)_6]^{2+} + CO_3^{2-} \rightarrow FeCO_3 + 6H_2O$$
$$\text{green ppt}$$

$[Cu(H_2O)_6]^{2+}$

The addition of sodium hydroxide solution or ammonia solution to a solution containing $[Cu(H_2O)_6]^{2+}$ ions produces a blue precipitate of copper(ii) hydroxide:

Exam tip

These ions are produced in solution when a compound containing the metal ion dissolves in water. For example, dissolving copper(ii) sulfate in water produces a solution containing $[Cu(H_2O)_6]^{2+}$ ions and SO_4^{2-} ions.

Exam tip

Remember that ppt is the abbreviation for precipitate. Iron(ii) hydroxide is often written as $Fe(OH)_2$, but in these equations you should include the water in the formula of the hydroxide ppt.

Knowledge check 15

What is observed when sodium hydroxide solution is added to iron(ii) sulfate solution?

$$[Cu(H_2O)_6]^{2+} + 2OH^- \rightarrow Cu(OH)_2(H_2O)_4 + 2H_2O$$
$$\text{blue ppt}$$

The precipitate is not soluble in excess sodium hydroxide solution but it does redissolve in excess ammonia solution, forming a deep blue solution. The equation for the redissolution of the blue precipitate in excess ammonia solution is:

$$Cu(OH)_2(H_2O)_4 + 4NH_3 \rightarrow [Cu(NH_3)_4(H_2O)_2]^{2+} + 2H_2O + 2OH^-$$
$$\text{deep blue solution}$$

When sodium carbonate solution is added to a solution containing $[Cu(H_2O)_6]^{2+}$ ions, a green precipitate of copper(II) carbonate is formed:

$$[Cu(H_2O)_6]^{2+} + CO_3^{2-} \rightarrow CuCO_3 + 6H_2O$$
$$\text{green ppt}$$

$[Fe(H_2O)_6]^{3+}$

When sodium hydroxide solution or ammonia solution is added to a solution containing $[Fe(H_2O)_6]^{3+}$ ions, a brown precipitate of iron(III) hydroxide is formed:

$$[Fe(H_2O)_6]^{3+} + 3OH^- \rightarrow Fe(OH)_3(H_2O)_3 + 3H_2O$$
$$\text{brown ppt}$$

The precipitate does not redissolve in either excess sodium hydroxide solution or excess ammonia solution.

When sodium carbonate solution is added to a solution containing $[Fe(H_2O)_6]^{3+}$ ions, a brown precipitate of iron(III) hydroxide is formed and bubbles of a gas are produced. The gas is carbon dioxide:

$$2[Fe(H_2O)_6]^{3+} + 3CO_3^{2-} \rightarrow 2Fe(OH)_3(H_2O)_3 + 3CO_2 + 3H_2O$$
$$\text{brown ppt}$$

$[Al(H_2O)_6]^{3+}$

When sodium hydroxide solution or ammonia solution is added to a solution containing $[Al(H_2O)_6]^{3+}$ ions, a white precipitate of aluminium hydroxide is formed:

$$[Al(H_2O)_6]^{3+} + 3OH^- \rightarrow Al(OH)_3(H_2O)_3 + 3H_2O$$
$$\text{white ppt}$$

The precipitate does not redissolve on addition of excess ammonia solution but it will redissolve on the addition of excess sodium hydroxide solution to form a colourless solution:

$$Al(OH)_3(H_2O)_3 + OH^- \rightarrow [Al(OH)_4(H_2O)_2]^- + H_2O$$
$$\text{colourless solution}$$

The OH^- ions in sodium hydroxide solution replace the water in the aluminium hydroxide precipitate. As more sodium hydroxide solution is added the replacement continues until $[Al(OH)_6]^{3-}$ is formed.

When sodium carbonate solution is added to a solution containing $[Al(H_2O)_6]^{3+}$ ions, a white precipitate of aluminium hydroxide is formed and bubbles of a gas are produced. The gas is carbon dioxide:

Knowledge check 16

What is observed when sodium carbonate solution is added to a solution of aluminium sulfate?

$$2[Al(H_2O)_6]^{3+} + 3CO_3^{2-} \rightarrow 2Al(OH)_3(H_2O)_3 + 3CO_2 + 3H_2O$$

$$\text{white ppt}$$

These reactions are summarised in Table 6.

Table 6

Hexaaqua complex	$[Fe(H_2O)_6]^{2+}$	$[Cu(H_2O)_6]^{2+}$	$[Fe(H_2O)_6]^{3+}$	$[Al(H_2O)_6]^{3+}$
Colour	Green	Blue	Yellow (or brown or violet or purple or lilac)	Colourless
Reaction with NaOH(aq)	Green ppt	Blue ppt	Brown ppt	White ppt
Complex formed	$Fe(OH)_2(H_2O)_4$	$Cu(OH)_2(H_2O)_4$	$Fe(OH)_3(H_2O)_3$	$Al(OH)_3(H_2O)_3$
Does it redissolve in excess NaOH(aq) (and colour of solution formed)?	No	No	No	Yes — colourless solution formed
Complex formed	–	–	–	$[Al(OH)_4(H_2O)_2]^-$ or $[Al(OH)_6]^{3-}$
Reaction with NH$_3$(aq)	Green ppt	Blue ppt	Brown ppt	White ppt
Complex formed	$Fe(OH)_2(H_2O)_4$	$Cu(OH)_2(H_2O)_4$	$Fe(OH)_3(H_2O)_3$	$Al(OH)_3(H_2O)_3$
Does it redissolve in excess NH$_3$(aq) (and colour of solution formed)?	No	Yes — deep blue solution formed	No	No
Complex formed	–	$[Cu(NH_3)_4(H_2O)_2]^{2+}$	–	–
Reaction with Na$_2$CO$_3$(aq)	Green ppt	Green ppt	Brown ppt; bubbles of gas released	White ppt; bubbles of gas released
Compound or complex formed	$FeCO_3$	$CuCO_3$	$Fe(OH)_3(H_2O)_3$	$Al(OH)_3(H_2O)_3$

Required practical 11

You will be required to carry out test tube reactions to identify transition metal ions in aqueous solution (using sodium hydroxide solution, ammonia solution and sodium carbonate solution).

Summary

- Transition metal complexes in solution react with sodium hydroxide solution, ammonia solution and sodium carbonate solution.
- Sodium hydroxide solution and ammonia solution react with transition metal ions complexes to form hydroxide precipitates.
- The hydroxide precipitates formed from $[Fe(H_2O)_6]^{2+}$ and $Fe(H_2O)_6]^{3+}$ do not redissolve in either sodium hydroxide solution or ammonia solution.
- The hydroxide precipitate formed from $[Al(H_2O)_6]^{3+}$ redissolves in sodium hydroxide solution but not in ammonia solution.
- The hydroxide precipitate formed from $[Cu(H_2O)_6]^{2+}$ redissolves in ammonia solution but not in sodium hydroxide solution.
- The addition of sodium carbonate solution to a solution containing a complex with 2+ ions results in the formation of a carbonate precipitate.
- The addition of sodium carbonate solution to a solution containing a complex with 3+ ions results in formation of a hydroxide precipitate and the evolution of carbon dioxide gas.

Organic chemistry

Optical isomerism

Molecules that have the same molecular formula but a different structural formula are known as structural isomers. **Stereoisomers** are molecules that have the same structural formula but a different arrangement of atoms in space. There are two types of stereoisomer — *E–Z* isomers, which you met in the second student guide of this series, and optical isomers.

Optical isomerism occurs in molecules that have a carbon atom with four different atoms or groups attached to it tetrahedrally. Optical isomers are asymmetric and have no centre, plane or axis of symmetry, hence they form two non-superimposable tetrahedral arrangements in space; one is the mirror image of the other.

Optical isomers (enantiomers) are stereoisomers that occur as a result of chirality in molecules. They exist as non-superimposable mirror images and differ in their effect on plane-polarised light.

When asked to draw optical isomers, a general method to follow is:

- Draw the displayed formula.
- Identify the chiral centre.
- Draw the three-dimensional tetrahedral structure based on the chiral centre and insert the four different groups.
- Draw a dotted line to represent a mirror, and draw the second isomer by either reflecting the isomer in an imaginary mirror (Figure 18a) or by simply exchanging two of the groups attached to the chiral atom (Figure 18b).

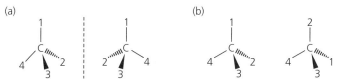

Figure 18 Optical isomers. The numbers 1–4 represent four different groups

Worked example 1

Draw the optical isomers of butan-2-ol.

Answer

First draw the displayed formula (Figure 19).

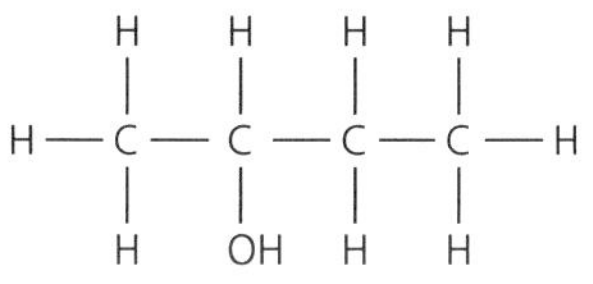

$$H-\overset{\overset{H}{|}}{\underset{\underset{H}{|}}{C}}-\overset{\overset{H}{|}}{\underset{\underset{OH}{|}}{C}}-\overset{\overset{H}{|}}{\underset{\underset{H}{|}}{C}}-\overset{\overset{H}{|}}{\underset{\underset{H}{|}}{C}}-H$$

Figure 19

An asymmetric carbon atom is chiral and has four different atoms or groups attached.

Chiral means that the structure and its image are non-superimposable.

Exam tip

Remember when drawing a three-dimensional tetrahedron that two of the bonds are in the plane of the page and are represented by lines. A wedged bond represents a bond coming out of the paper towards the viewer, and a dashed bond represents a bond going into the paper.

Then identify the chiral centre. The two end carbon atoms have three hydrogen atoms bonded to them and the third carbon atom from the left has two hydrogen atoms bonded to it, so these cannot be chiral centres. The second carbon atom from the left has the following groups bonded to it:

1 CH_3 **2** OH   **3** H   **4** CH_2CH_3

These are four different groups, so it is a chiral centre and is marked with an asterisk (Figure 20).

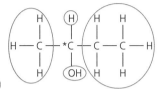

Figure 20

Draw a three-dimensional tetrahedral arrangement and insert each of the four different groups at different points on the tetrahedron. Then place a dotted line to represent the mirror, and reflect the image as shown in Figure 21.

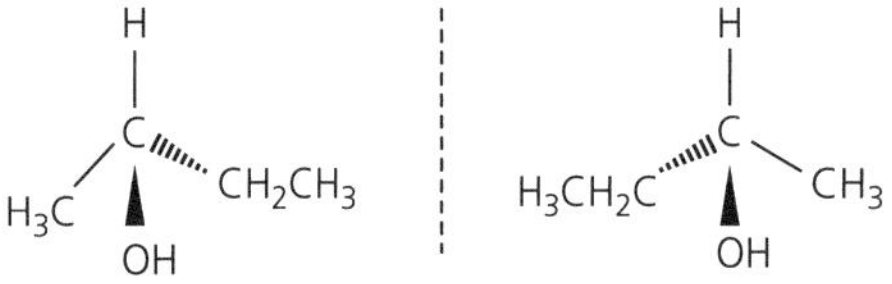

Figure 21

Alternatively, to draw the optical isomers you can exchange any two of the groups attached to the chiral centre. An example is shown in Figure 22.

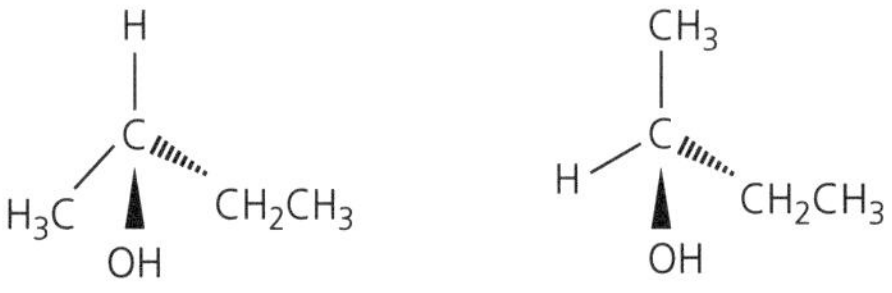

Figure 22

These isomers cannot be superimposed on each other. They have the same molecular and structural formula and differ only in the arrangement of groups around the chiral centre. You are expected to be able to draw optical isomers for molecules that have a single chiral centre.

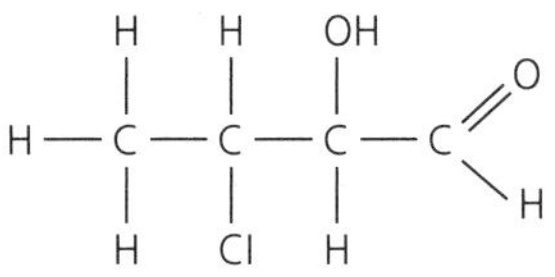

Optical activity

An **optically active substance** is one that can rotate the plane of plane-polarised light. **Plane-polarised light** is light in which all the waves vibrate in the same plane. Optical isomers each rotate the plane of plane-polarised light in opposite directions and hence they are optically active.

Mixing equal amounts of the same concentration of two enantiomers gives an **optically inactive mixture**, which has no effect on plane-polarised light because the two opposite effects cancel out. This mixture of equal amounts of each enantiomer is called a **racemic mixture** or **racemate.**

Exam tip

Circle each of the four groups on the chiral carbon — this helps you to remember which groups to place around the tetrahedron.

Knowledge check 17

Identify any chiral centres present in this molecule:

An **optically active substance** can rotate the plane of plane-polarised light.

Plane polarised light is light in which all the waves vibrate in the same plane.

A **racemic mixture** (racemate) has equal amounts of enantiomers.

Worked example

Explain how you could distinguish between a racemate of alanine and one of the enantiomers of alanine?

Answer

Pass plane-polarised light into the two solutions.

Plane-polarised light will be rotated by the single enantiomer but it will be unaffected by the racemate.

The racemate is optically inactive because it contains equal amounts of each isomer. One isomer rotates plane-polarised light to the right, the other to the left, and the two opposite effects cancel out.

Knowledge check 18

Name an isomer of C_4H_9OH that can exist as optical isomers and state how a solution of one of the optical isomers can be distinguished from the other.

Summary

- Optical isomers are a type of stereoisomer formed as a result of chirality in molecules. An asymmetric carbon atom is chiral and has four different atoms or groups attached.
- Optical isomers exist as non-superimposable mirror images and each optical isomer (enantiomer) rotates the plane of plane-polarised light in a different direction.
- A mixture of equal amounts of enantiomers is a racemate and is optically inactive because one isomer rotates plane-polarised light to the left, and the other to the right, and the two opposite effects cancel.

Aldehydes and ketones

Aldehydes and ketones both contain the carbonyl group, which is polar:

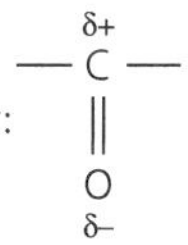

$$\begin{array}{c} \overset{\delta+}{-C-} \\ \parallel \\ O \\ \delta- \end{array}$$

Nomenclature of aldehydes and ketones

The names of aldehydes are based on the carbon skeleton, with the ending changed to -anal (Figures 23 and 24). The carbonyl group is always at the end of the chain and so a positional number is not needed.

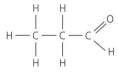

Methanal   Ethanal   Propanal

Figure 23

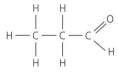

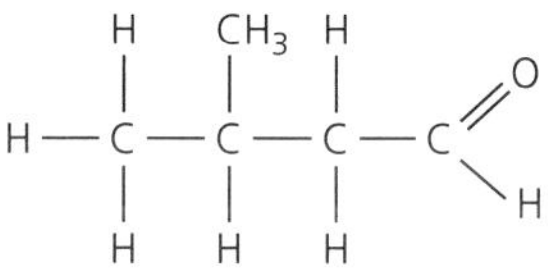

Figure 24

Exam tip

The name of the structure in Figure 24 is 3-methylbutanal. It is an aldehyde because C=O is at the end of the chain in position 1; the methyl group is in position 3.

The names of ketones are based on the carbon skeleton, with the ending changed to -anone (Figure 25). The carbonyl group can be at any position on the chain, except for the end.

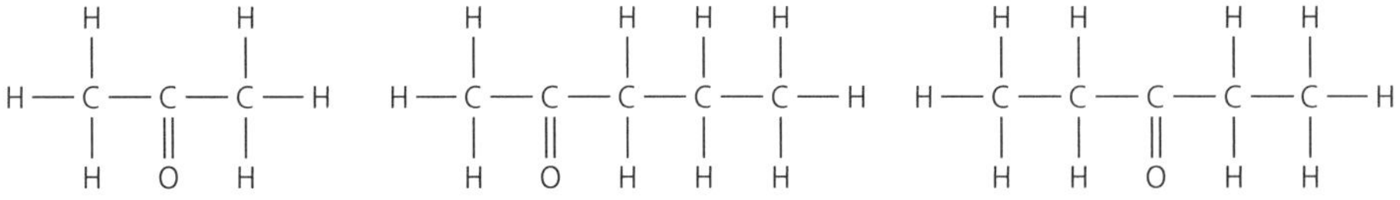

Propanone   Pentan-2-one   Pentan-3-one

Figure 25

Oxidation

Aldehydes and ketones are the products of the oxidation of alcohols. Aldehydes can be oxidised into carboxylic acids, and ketones cannot be oxidised. It is useful to remember the following sequences:

primary alcohol $\xrightarrow{[O]}$ aldehyde $\xrightarrow{[O]}$ carboxylic acid

secondary alcohol $\xrightarrow{[O]}$ ketone $\xrightarrow{\hspace{0.3cm}}\!\!\times\!\!\xrightarrow{\hspace{0.3cm}}$ no further oxidation

The oxidising agent is a solution of acidified potassium dichromate(VI), which is represented by [O]. The *orange* dichromate ion is reduced to the *green* chromium(III) ion, Cr^{3+}, by the aldehyde according to the ionic equation:

$$Cr_2O_7{}^{2-} + 14H^+ + 6e^- \rightarrow 2Cr^{3+} + 7H_2O$$

Example: Oxidation of an aldehyde

$$CH_3CHO \qquad + [O] \quad \rightarrow \qquad CH_3COOH$$
ethanal (aldehyde)   ethanoic acid (carboxylic acid)

Conditions: warm with acidified potassium dichromate(VI)

Observation: orange solution changes to green solution

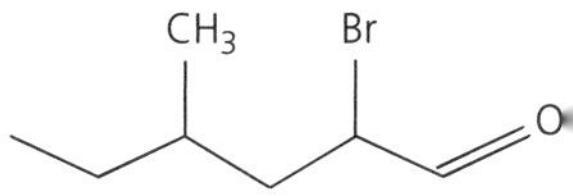

Chemical tests to distinguish between aldehydes and ketones

Fehling's solution and Tollens' reagent are mild oxidising agents that are used to distinguish between aldehydes and ketones (Table 7). Aldehydes are oxidised and ketones are not.

Table 7

Solution	Method	Result	Equation
Fehling's solution	Add a few drops of the unknown solution to 1 cm³ of freshly prepared Fehling's solution reagent in a test tube. Warm in a water bath.	If the unknown is an aldehyde an orange-red ppt occurs. Solution remains blue for a ketone.	$Cu^{2+} + e^- \rightarrow Cu^+$
Tollens' reagent	Add a few drops of the unknown solution to 1 cm³ of freshly prepared Tollens' reagent. Warm in a water bath.	If the unknown is an aldehyde a silver mirror occurs on the test tube. Solution remains colourless for a ketone.	$Ag^+ + e^- \rightarrow Ag$

Exam tip

Ketones require a minimum of three carbon atoms as there must be a C=O in the chain, not at the end. Propanone is the simplest ketone. Ketones require a number for the position of the CO group from five carbon atoms upwards.

Knowledge check 19

Name this structure:

Nucleophilic addition reactions

The carbonyl group is unsaturated and can undergo addition reactions. The carbonyl group is also polar and the carbon δ^+ is susceptible to attack by nucleophiles. Hence aldehydes and ketones take part in nucleophilic addition reactions.

Reduction of aldehydes and ketones using sodium tetrahydridoborate(III)

Sodium tetrahydridoborate(III), $NaBH_4$, is a reducing agent. It is represented by [H] in equations.

Equation: $CH_3CH_2CHO + 2[H] \rightarrow CH_3CH_2CH_2OH$
         propanal   propan-1-ol

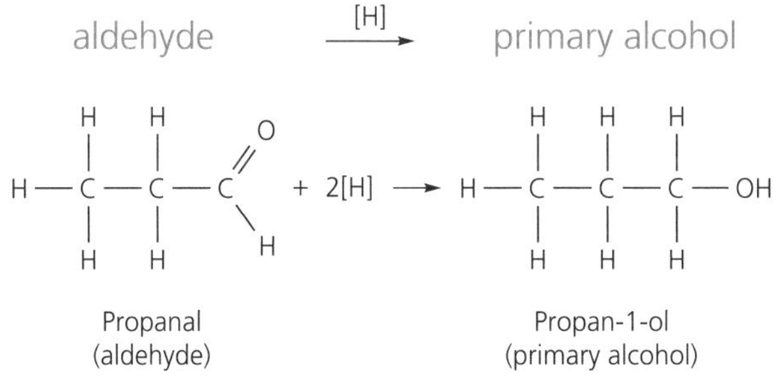

aldehyde $\xrightarrow{[H]}$ primary alcohol

Propanal
(aldehyde)

Propan-1-ol
(primary alcohol)

Conditions: heat under reflux with sodium tetrahydridoborate(III) in aqueous ethanol followed by acidification with dilute sulfuric acid.

Equation: $CH_3COCH_3 + 2[H] \rightarrow CH_3CH(OH)CH_3$

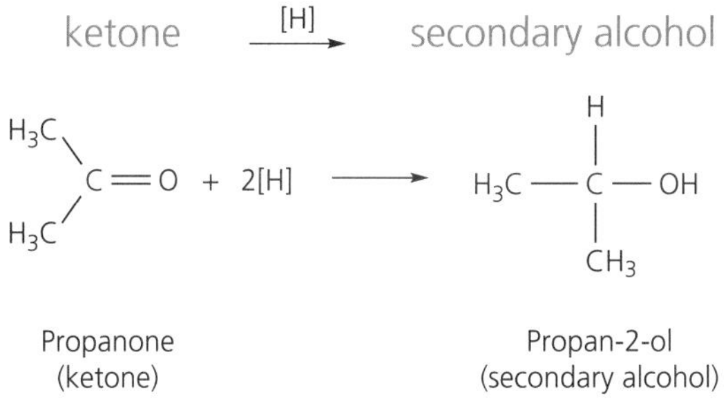

ketone $\xrightarrow{[H]}$ secondary alcohol

Propanone
(ketone)

Propan-2-ol
(secondary alcohol)

Conditions: heat under reflux with sodium tetrahydridoborate(III) in **aqueous solution**, followed by acidification with dilute sulfuric acid.

Mechanism: the mechanism for this reaction is nucleophilic addition. The BH_4^- ion in $NaBH_4$ is a source of hydride ions (H^-). The hydride ion acts as a nucleophile and attacks the carbon $\delta+$. The mechanism for propanone and $NaBH_4$ is shown in Figure 26.

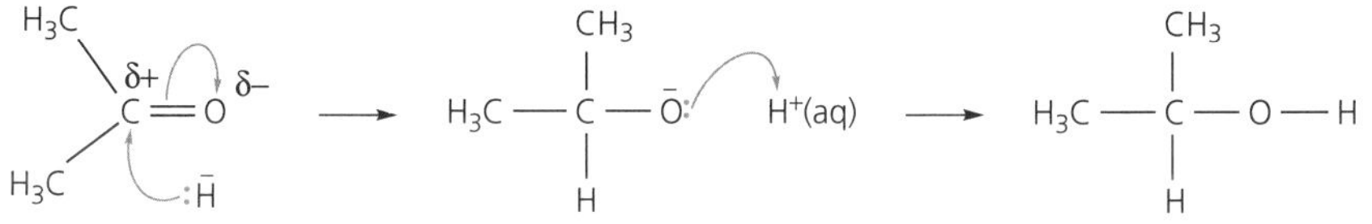

Figure 26

An addition reaction is one in which the π bond of the double bond is broken and species are added across the double bond.

A nucleophile is a lone pair donor. It is an atom or group that is attracted to an electron-deficient centre, where it donates a lone pair to form a new covalent bond.

Reaction of aldehydes and ketones with KCN followed by dilute acid

In this reaction a hydroxynitrile is produced.

Equation:

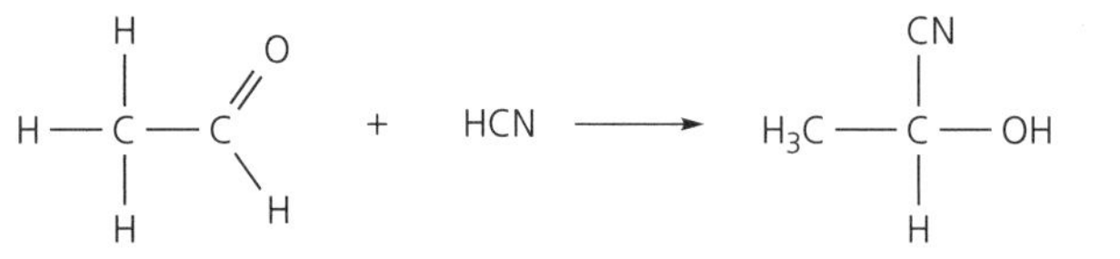

Ethanal
(aldehyde)

2-hydroxypropanenitrile
(hydroxynitrile)

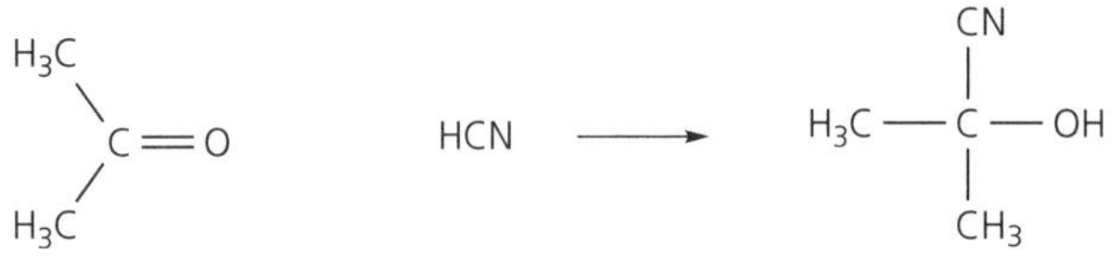

Propanone
(ketone)

2-hydroxy-2-methylpropanenitrile

Conditions: add dilute acid to an aqueous solution of potassium cyanide to generate hydrogen cyanide (HCN) in the reaction mixture.

Mechanism: the mechanism is nucleophilic addition. The cyanide ion is the nucleophile. The mechanism for ethanal and KCN and dilute acid is shown in Figure 27.

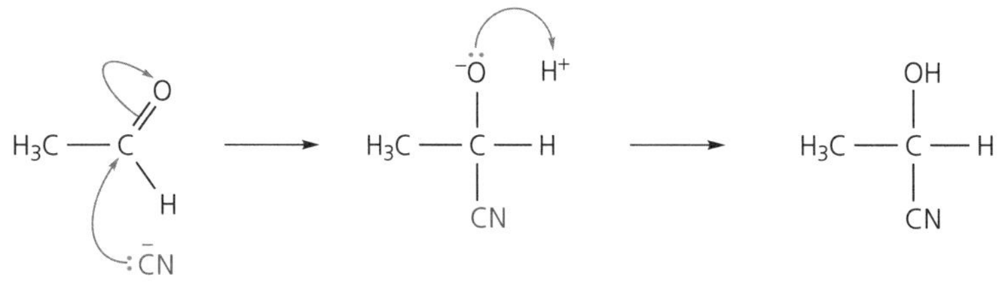

Figure 27

Aldehydes and unsymmetrical ketones form mixtures of enantiomers when they react with KCN followed by dilute acid. In the reaction of ethanal with potassium cyanide to produce 2-hydroxypropanenitrile the product is optically inactive because a racemic mixture — a 50/50 mixture of the two optical isomers — is formed.

All aldehydes produce a racemate in this reaction.

Unsymmetrical ketones, for example $CH_3COCH_2CH_3$, will produce a racemate.

Symmetrical ketones, for example CH_3COCH_3, produce a product that does not have an asymmetric carbon and is optically inactive.

Exam tip

When naming hydroxynitriles, the carbon with nitrogen attached is always counted as the first carbon in the chain.

Exam tip

Extreme care must be used when handling potassium cyanide as it is toxic when ingested and forms hydrogen cyanide, an extremely toxic gas, when in contact with acid.

Exam tip

Remember that in the formation of a covalent bond the curly arrow starts from a lone pair or from another covalent bond. In the breaking of a covalent bond the curly arrow starts from the bond.

Summary

- Aldehydes and ketones both contain the carbonyl group. Aldehydes can be oxidised by acidified potassium dichromate(VI) to carboxylic acids. Ketones cannot be oxidised.
- Aldehydes give a silver mirror with Tollens' reagent and a red ppt with Fehling's solution. Ketones do not react. This is a test to distinguish between aldehydes and ketones.
- Aldehydes can be reduced to primary alcohols, and ketones to secondary alcohols, using $NaBH_4$ in aqueous solution. These reduction reactions are nucleophilic addition reactions.
- Carbonyl compounds react with KCN, followed by dilute acid, to produce hydroxynitriles in a nucleophilic addition reaction.
- Aldehydes and unsymmetrical ketones form mixtures of enantiomers when they react with KCN followed by dilute acid.

Carboxylic acids and derivatives

Carboxylic acids

Naming carboxylic acids

Carboxylic acids are named according to IUPAC rules. The names are based on the carbon skeleton, with the ending changed from –ane to –anoic acid.

IUPAC nomenclature rules state that *the carboxyl carbon in the COOH functional group is always carbon number 1.* Any substituents are numbered according to this. For example, the structure in Figure 28 is 3-hydroxy-3-methylhexanoic acid.

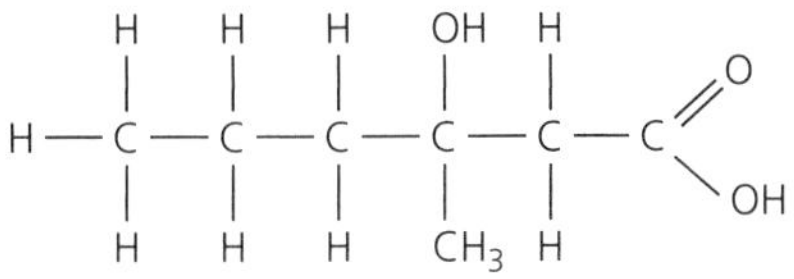

Figure 28 3-hydroxy-3-methylhexanoic acid

Acid reactions of carboxylic acids

All carboxylic acids in aqueous solution act as acids, dissociating to form $H^+(aq)$ (or $H_3O^+(aq)$) and the carboxylate ion. They are **weak acids** because they are partially dissociated in solution. For example:

$CH_3COOH(aq) \rightarrow CH_3COO^-(aq) + H^+_{(aq)}$
ethanoic acid   ethanoate ion   hydrogen ion

or

$CH_3COOH + H_2O \rightarrow CH_3COO^- + H_3O^+$
ethanoic acid water   ethanoate ion   hydroxonium ion

Carboxylic acids take part in typical acid reactions — with carbonates, metals and bases to form salts.

Exam tip

Carboxylic acids have the structure:

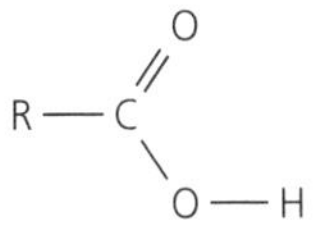

where R is an alkyl group. The functional group is the **carboxyl** group –COOH. It contains a carbonyl and a hydroxyl group and is drawn:

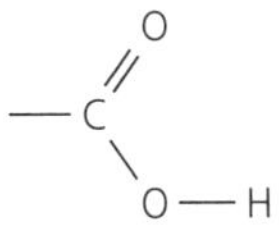

Exam tip

Carboxylic acids are soluble in water because the highly polar carbonyl and hydroxyl groups can hydrogen bond with water.

With carbonates

Equation: **acid + carbonate → salt + water + carbon dioxide**

For example:

$$2CH_3COOH + Na_2CO_3 \rightarrow 2CH_3COONa + CO_2 + H_2O$$
ethanoic acid   sodium carbonate   sodium ethanoate

Observations: there will be effervescence and the solid sodium carbonate will be used up, producing a colourless solution.

An example with a hydrogen carbonate is:

$$CH_3COOH + NaHCO_3 \rightarrow CH_3COONa + CO_2 + H_2O$$
ethanoic acid   sodium   sodium ethanoate
                 hydrogencarbonate

Test for a carboxylic acid: despite being weak acids, carboxylic acids are stronger than carbonic acid and release carbon dioxide, which changes colourless limewater cloudy, when reacted with carbonates. This is the reaction used to test for carboxylic acids.

With metals

Equation: **acid + metal → salt + hydrogen**

For example:

$$2CH_3COOH + Mg \rightarrow (CH_3COO)_2Mg + H_2$$
ethanoic acid   magnesium ethanoate

Observation: there will be effervescence and the solid magnesium will be used up, producing a colourless solution.

With bases

Equation: **acid + base → salt + water**

For example:

$$CH_3COOH + NaOH \rightarrow CH_3COONa + H_2O$$
ethanoic acid   sodium hydroxide   sodium ethanoate

Observations: there is release of heat and the colourless solution remains.

For the reaction of a carboxylic acid with the base ammonia, only an ammonium salt is produced:

Equation: $CH_3COOH + NH_3 \rightarrow CH_3COONH_4$
              ethanoic acid   ammonia   ammonium ethanoate

Observations: there is release of heat and the colourless solution remains.

Esters

Carboxylic acids react with alcohols in the presence of a strong acid catalyst, to produce esters. Esters are carboxylic acid derivatives and have the general structure:

$$R_1 - \underset{\underset{O}{\|}}{C} - O - R_2$$

R_1 is from the acid and R_2 is from the alcohol.

> **Exam tip**
>
> The formula of sodium ethanoate can be written $CH_3COO^-Na^+$, to stress the ionic nature of the salt. It is incorrect to put one charge in and omit the other.

> **Knowledge check 20**
>
> Name the products and write the equation when magnesium oxide reacts with propanoic acid.

The formation of an ester can be represented by the equation:

$$R_1 - \overset{\displaystyle O}{\underset{\displaystyle O-H}{C}} \quad + \quad R_2OH \quad \rightleftharpoons \quad R_1 - \overset{\displaystyle O}{\underset{\displaystyle O-R_2}{C}} \quad + \quad H_2O$$

> **Exam tip**
>
> The functional group of an ester is an ester group or ester linkage $-COO-$, drawn as:
>
> $$-\overset{}{\underset{\displaystyle O}{C}} - O -$$

Naming esters

An ester is an **alkyl carboxylate**. When naming, the alcohol provides the **alkyl** part of the name and the carboxylic acid provides the **carboxylate** part of the name. For example, the ester made from methanol and propanoic acid is methyl propanoate, $CH_3CH_2COOCH_3$, and the ester made from butanoic acid and propanol is propyl butanoate, $CH_3CH_2CH_2COOCH_2CH_2CH_3$.

> **Exam tip**
>
> It is best when drawing the structural formula of the ester to start with the acid end of the molecule.
>
> Acid alkyl ——— $\overset{}{\underset{\displaystyle O}{C}}$ — O ——— Alcohol alkyl
> group   group

> **Knowledge check 21**
>
> Name and write the formula of the ester made from propanoic acid and ethanol.

Esterification equations

Carboxylic acids react with alcohols to produce esters, in an equilibrium reaction. For example:

HCOOH	+	$CH_3CH_2CH_2OH$	$\rightleftharpoons$	$HCOOCH_2CH_2CH_3$	+	H_2O
methanoic acid		propan-1-ol		propyl methanoate		water

$$H - \overset{\displaystyle O}{\underset{\displaystyle O-H}{C}} \quad + \quad H-\overset{\displaystyle H}{\underset{\displaystyle H}{C}}-\overset{\displaystyle H}{\underset{\displaystyle H}{C}}-\overset{\displaystyle H}{\underset{\displaystyle H}{C}}-O-H \quad \rightleftharpoons \quad H-\overset{}{\underset{\displaystyle O}{C}}-O-\overset{\displaystyle H}{\underset{\displaystyle H}{C}}-\overset{\displaystyle H}{\underset{\displaystyle H}{C}}-\overset{\displaystyle H}{\underset{\displaystyle H}{C}}-H \quad + \quad H_2O$$

Conditions: a catalyst of concentrated sulfuric acid is used and the mixture is heated.

Uses of esters

- **Plasticisers** — additives mixed into polymers to improve their flexibility are often esters.
- Esters such as ethyl ethanoate are often used as **solvents** in paints.
- **Perfumes** contain many sweet-smelling esters.
- **Food flavourings** are often esters, for example pentyl pentanoate gives a pineapple flavour.

> **Exam tip**
>
> Questions may be set that give the structure of the ester and expect you to determine the structure of the acid and alcohol from which they are formed.

Preparation of a liquid organic substance (ester)

The ester ethyl ethanoate is prepared from ethanol and ethanoic acid, using the method given below:

1 Mix equimolar volumes of ethanoic acid and ethanol in a pear-shaped flask.

2 Add $1\,cm^3$ of concentrated sulfuric acid *slowly, with cooling and shaking,* to prevent splashing as the mixture gets hot.

3 Add some antibumping granules and heat under reflux for 20 minutes.

4 Distil off the ethyl ethanoate and collect the fraction around the boiling point temperature.

5 Place the crude ethyl ethanoate in a separating funnel and shake with sodium carbonate solution to remove traces of unreacted ethanoic acid and the concentrated sulfuric acid. Invert the funnel and open the tap occasionally to release pressure due to any carbon dioxide gas produced.

6 Allow the layers to separate and discard the lower aqueous layer.

7 Add some calcium chloride solution to the ethyl ethanoate, to remove any ethanol impurities, shake and run off the lower aqueous layer.

8 Place ester in a boiling tube.

9 Add a spatula of anhydrous calcium chloride (a drying agent that removes water) and stopper the boiling tube and shake. Repeat until the cloudy ester changes to clear as the water is removed.

10 Filter to remove calcium chloride.

11 Redistill to remove any remaining organic impurities, collecting the fraction at the boiling point.

> ### Required practical 10
>
> You will be required to prepare a sample of a pure organic liquid (preparation of a liquid ester).

Hydrolysis of esters

Hydrolysis is the reverse of an esterification reaction. The ester is split (lysis) by the action of water (hydro) into the carboxylic acid and alcohol. The hydrolysis of an ester needs heat and occurs in either dilute mineral acid such as hydrochloric acid or a solution of an alkali such as sodium hydroxide.

Acid hydrolysis

In acidic conditions, esters are not completely hydrolysed — an equilibrium mixture is formed in which some ester is present.

Equation: **ester + water ⇌ carboxylic acid + alcohol**

For example:

$$CH_3COOC_2H_5 \;+\; H_2O \;\rightleftharpoons\; CH_3COOH \;+\; C_2H_5OH$$
ethyl ethanoate   water   ethanoic acid   ethanol

Conditions: heat under reflux with dilute sulfuric or hydrochloric acid.

Exam tip

Antibumping granules promote smooth boiling.

Exam tip

Remember that reflux is the continuous boiling and condensing of a reaction mixture. Refer to the second student guide in this series to refresh your memory on reflux, distillation and the use of a separating funnel.

Exam tip

Esters can be hydrolysed in acid or alkaline conditions to form alcohols and carboxylic acids or salts of carboxylic acids.

Alkaline hydrolysis

In alkaline conditions, esters undergo complete hydrolysis, forming the corresponding alcohol and the salt of the carboxylic acid. The reaction in alkaline solution is quicker.

Equation: **ester + water → carboxylic acid salt + alcohol**

For example:

$$CH_3COOC_2H_5 \quad + \quad NaOH \quad \rightarrow \quad CH_3COONa \quad + \quad C_2H_5OH$$

ethyl ethanoate   sodium hydroxide   sodium ethanoate   ethanol

Conditions: heat under reflux with aqueous sodium hydroxide.

In order to liberate the free acid from its salt in alkaline hydrolysis, a dilute mineral acid such as dilute hydrochloric acid should be added:

$$CH_3COONa + HCl \rightarrow CH_3COOH + NaCl$$

This alkaline hydrolysis is sometimes called saponification and is the basis of soap making.

Vegetable oils and fats

Vegetable oils and animal fats are **esters** of propane-1,2,3-triol (a tri-alcohol) and a long-chain carboxylic acid called a fatty acid.

Propane-1,2,3-triol (also known as **glycerol**) has the structure shown in Figure 29.

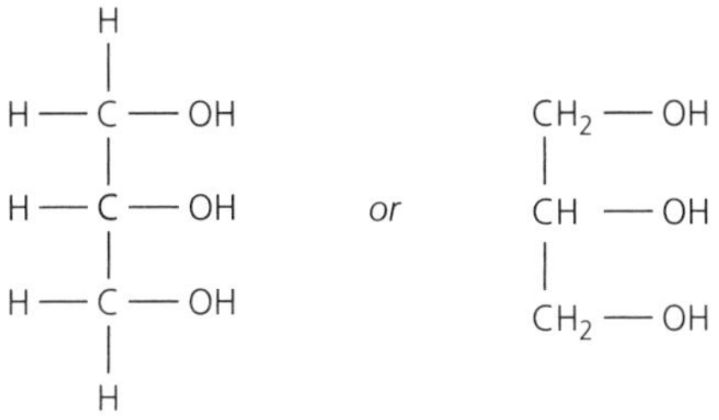

Figure 29

Fatty acids generally have the structure shown in Figure 30. The R is hydrogen or an alkyl group.

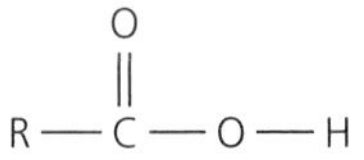

Figure 30

Two common fatty acids are stearic acid, $CH_3(CH_2)_{16}COOH$, a saturated fatty acid, and oleic acid $CH_3(CH_2)_7CH=CH(CH_2)_7COOH$, an unsaturated fatty acid.

The fats formed in the condensation reaction between fatty acids and glycerol are triesters and are often referred to as triglycerides. Propane-1,2,3-triol has three OH groups and so it reacts with three fatty acids to form a triglyceride (Figure 31).

Exam tip

It is important to remember that alkaline hydrolysis produces the sodium salt of the carboxylic acid and the alcohol, whereas acid hydrolysis produces the carboxylic acid and the alcohol.

Fatty acids are naturally occurring long-chain carboxylic acids.

Saturated fatty acids are fatty acids that do not have a double bond in the hydrocarbon chain.

Unsaturated fatty acids are fatty acids that have at least one C=C double bond in the hydrocarbon chain.

Triglycerides are esters of propane-1,2,3-triol (glycerol) and three fatty acid molecules. The fatty acids may or may not be all the same.

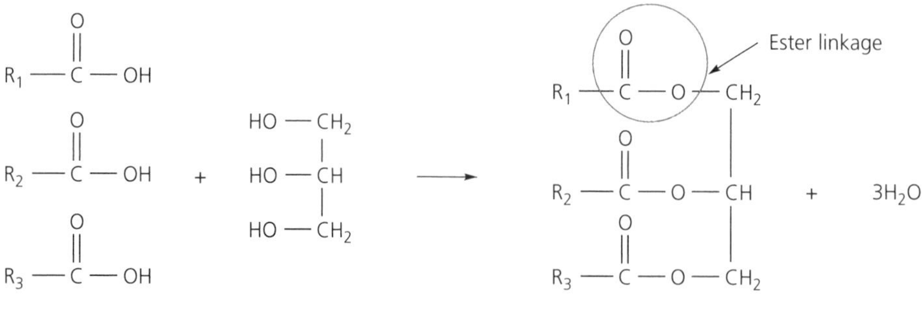

Three fatty acids Glycerol Triglyceride ester

Figure 31

It is important to note the triester linkage in the triglyceride. The three fatty acids that form the lipid may be the same (e.g. three stearic acids) or they may be different.

Hydrolysis of vegetable oils and fats to make soap

Fats and oils are esters and can be hydrolysed using hot alkali in the process of **saponification** (Figure 32).

Saponification is the alkaline hydrolysis of fats into glycerol and the salts of the fatty acids present in the fat.

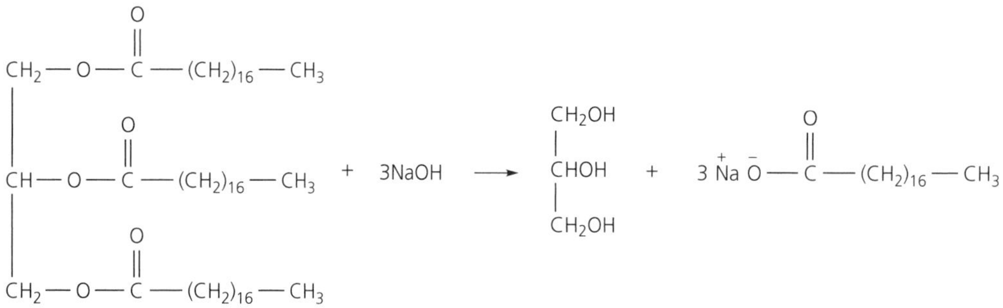

Figure 32 Saponification

Conditions: heat under reflux with aqueous sodium hydroxide.

The sodium or potassium salts of the fatty acids which are formed are called **soaps**.

Biodiesel

Biodiesel is a renewable fuel that consists of a mixture of methyl esters of long-chain carboxylic acids (fatty acids). It is produced by heating vegetable oils (triglycerides) with methanol in the presence of an acid catalyst. The process can be called **trans-esterification** — reacting a ester with an alcohol to produce a different ester and a different alcohol (Figure 33).

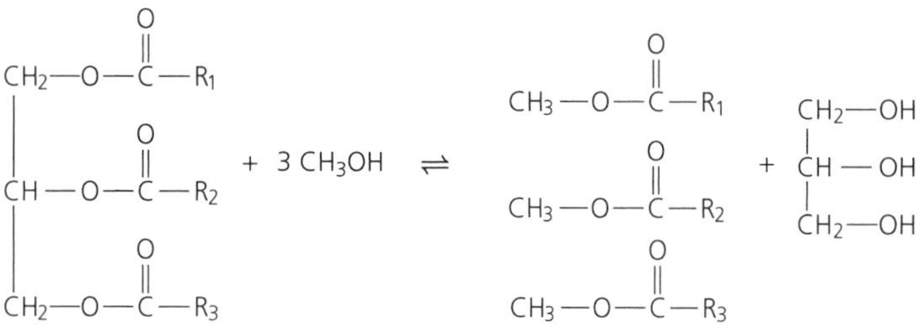

Triglyceride Methanol Mixture of methyl esters of fatty acids (biodiesel) Glycerol

Figure 33 Trans-esterification

The alkyl groups R_1, R_2 and R_3 can be the same or different. The reaction is reversible, so an excess of methanol is used to drive the equilibrium to the right. Under appropriate conditions this process can produce a 98% yield.

Knowledge check 22

Canola oil has the structure shown in Figure 34. Write the formula of the biodiesel formed from this oil and methanol.

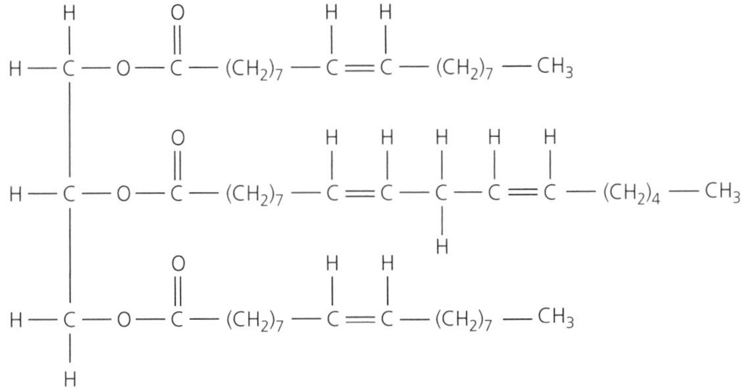

Figure 34

Acylation reactions of carboxylic acid derivatives

Acid derivatives such as acyl chlorides, acid anhydrides and amides are compounds that are related to carboxylic acids; the OH group has been replaced by something else (Table 8).

Table 8

Acyl chlorides	Acid anhydride	Amide
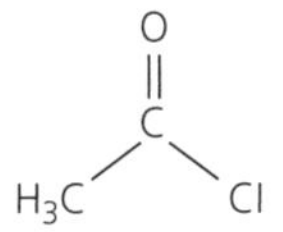 Ethanoyl chloride	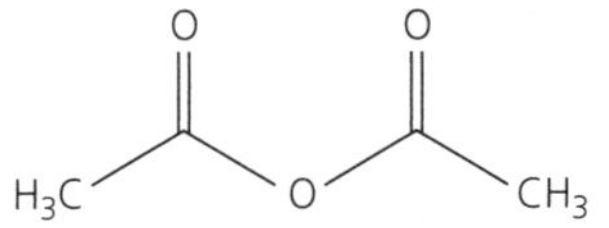 Ethanoic anhydride, $(CH_3CO)_2O$	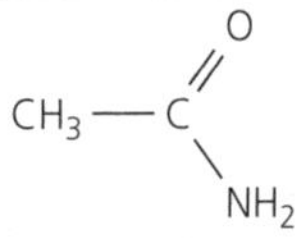 Ethanamide
The OH of the acid is replaced by a chlorine atom	The OH is replaced by $OCOCH_3$ when two carboxylic acids join and water is eliminated	The OH of the acid is replaced by an NH_2 group

An acyl functional group has the structure shown in Figure 35.

Acylation can be carried out using acid derivatives such as acyl chlorides and acid anhydrides, which act as acylating agents. In general, an acylating agent can be represented by the structure shown in Figure 36.

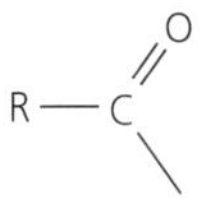

Figure 35 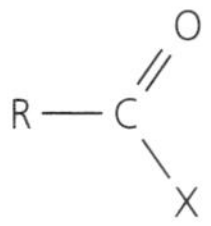   Figure 36

In acid chlorides X = Cl; in anhydrides X = OCOR.

> **Acylation** is the process of replacing a hydrogen atom in certain molecules by an acyl group (RCO–).

> ### Exam tip
> Notice that in these reactions the organic product is the same. The only difference is the other products formed — acyl chlorides form HCl and acid anhydrides form CH_3COOH.

Acylation of alcohols

With acyl chloride

Equation: $CH_3COCl + CH_3CH_2OH \rightarrow CH_3COOCH_2CH_3 + HCl$
      ethanoyl chloride   ethanol   ethyl ethanoate   hydrogen chloride

Observation: a vigorous reaction that produces steamy fumes of HCl(g); heat.

This is a suitable way of producing an **ester** from an alcohol because it occurs at room temperature, is irreversible and the **hydrogen chloride** is removed as a gas, shifting the equilibrium to the right and forming more of the ester. This method is not commonly used in the laboratory due to the volatile and poisonous nature of the acid chlorides. The normal laboratory preparation of an ester uses an alcohol and a carboxylic acid and needs heat, a catalyst and is reversible, so it is more difficult to get a high yield of ester.

With acid anhydride

Equation: $(CH_3CO)_2O + CH_3CH_2OH \rightarrow CH_3COOCH_2CH_3 + CH_3COOH$
      ethanoic anhydride   ethanol   ethyl ethanoate   ethanoic acid

Observation: this is a slower and less vigorous reaction than that of an acid chloride and water. The reaction needs warming.

An **ester** is produced in addition to **ethanoic acid**. Commercially, acid anhydrides are used preferentially to acyl chlorides in acylation reactions, as the reactions are easier to control.

The mechanism is nucleophilic addition–elimination (Figure 37).

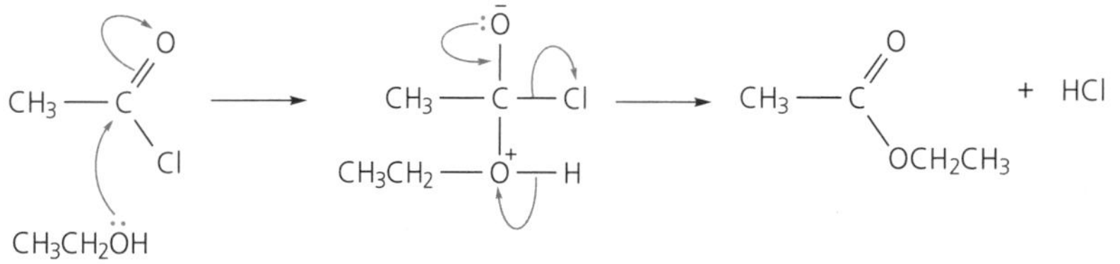

Figure 37 Nucleophilic addition–elimination — reaction of ethanoyl chloride with ethanol

Acylation of water

With acyl chloride

Equation: $R\text{–}COCl + H_2O \rightarrow R\text{–}COOH + HCl$

   $CH_3CH_2COCl + H_2O \rightarrow CH_3CH_2COOH + HCl$
   propanoyl chloride   propanoic acid   hydrogen chloride

Observations: vigorous reaction producing steamy fumes of hydrogen chloride.

With acid anhydride

Equation: $R_1\text{–}COOCO\text{–}R_2 + H_2O \rightarrow R_1\text{–}COOH + R_2\text{–}COOH$

In this case, two molecules of carboxylic acid are produced:

$$(CH_3CH_2CO)_2O + H_2O \rightarrow 2CH_3CH_2COOH$$

propanoic anhydride   propanoic acid

Observations: a slower reaction at room temperature. Two colourless solutions produce another colourless solution.

The mechanism is nucleophilic addition–elimination (Figure 38).

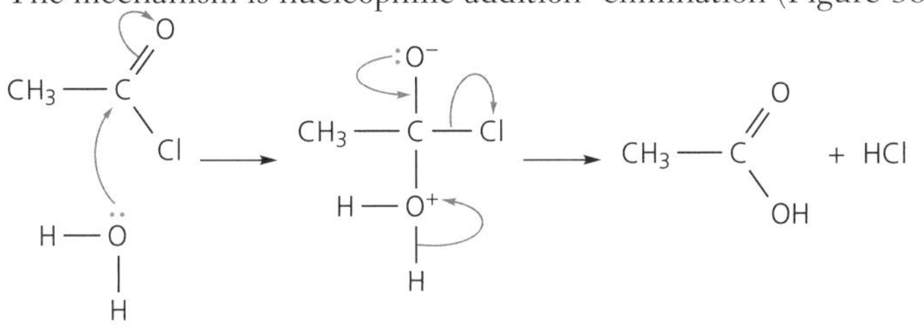

Figure 38 Nucleophilic addition–elimination — reaction of ethanoyl chloride with water

Acylation of ammonia

With acyl chloride

First an **amide** and hydrogen chloride are formed and then the basic ammonia reacts with the hydrogen chloride to form an ammonium salt.

Equation: $CH_3COCl + NH_3 \rightarrow CH_3CONH_2 + HCl$

$$NH_3 + HCl \rightarrow NH_4Cl$$

As a result, two ammonia molecules react:

$$CH_3COCl \quad + \quad 2NH_3 \quad \rightarrow \quad CH_3CONH_2 + \quad NH_4Cl$$

ethanoyl chloride   ethanamide   ammonium chloride

Observations: a violent reaction producing white smoke, which is a mixture of solid ammonium chloride and ethanamide. Some of the mixture remains dissolved in water as a colourless solution.

With acid anhydride

First an amide and hydrogen chloride are formed and the basic ammonia then reacts with the hydrogen chloride to form an ammonium salt:

Equation: $(CH_3CO)_2O + NH_3 \rightarrow CH_3CONH_2 + CH_3COOH$

$$NH_3 + CH_3COOH \rightarrow CH_3COONH_4$$

As a result, two ammonia molecules react:

$$(CH_3CO)_2O \quad + \quad 2NH_3 \quad \rightarrow \quad CH_3CONH_2 \quad + \quad CH_3COONH_4$$

ethanoic anhydride   ethanamide   ammonium ethanoate

Observation: a slower reaction than that of ethanol and ethanoyl chloride. Heating may be needed.

The mechanism is nucleophilic addition–elimination (Figure 39).

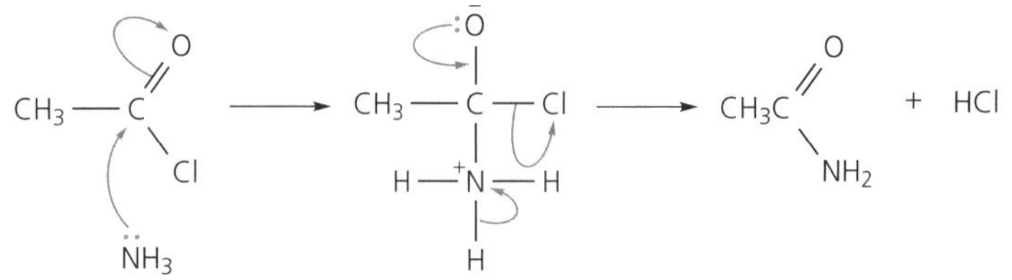

Figure 39 Nucleophilic addition–elimination — reaction of ethanoyl chloride with ammonia

Acylation of primary amines

With acyl chloride

An **N-substituted amide** is formed — this means that an alkyl group has substituted one of the hydrogen atoms of the NH_2 group. The reaction occurs in two stages. First, the N-substituted amine and hydrogen chloride form, but the amine is basic and reacts with the hydrogen chloride forming the ammonium salt, ethylammonium chloride. This is shown in Figure 40.

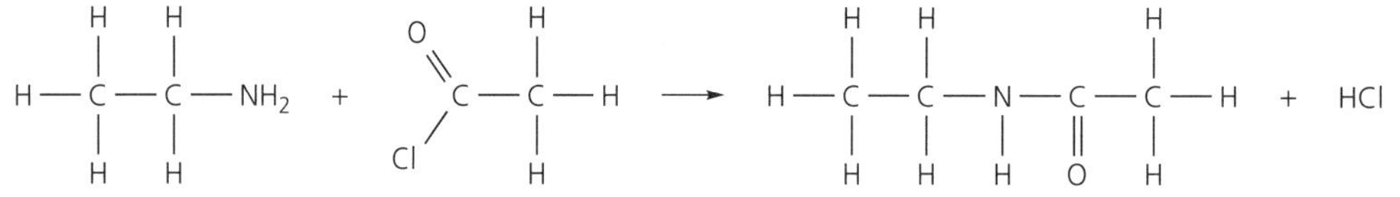

| Ethylamine | Ethanoyl chloride | N-ethylethanamide |

$$CH_3CH_2NH_2 + HCl \rightarrow CH_3CH_2NH_3Cl$$

Figure 40 Reaction of ethanoyl chloride and ethylamine

As a result, two amine molecules react:

$$CH_3COCl + 2CH_3CH_2NH_2 \rightarrow CH_3CONHCH_2CH_3 + CH_3CH_2NH_3Cl$$
ethanoyl chloride ethylamine   N-ethylethanamide ethylammonium chloride

With acid anhydride

$$(CH_3CH_2CO)_2O + CH_3NH_2 \rightarrow CH_3CH_2CONHCH_3 + CH_3COOH$$
propanoic anhydride methylamine N-methylpropanamide ethanoic acid

The amine is basic and will react with the ethanoic acid formed to produce a salt:

$$CH_3COOH + CH_3NH_2 \rightarrow CH_3COONH_3CH_3$$

The salt is called methylammonium ethanoate. It is just like ammonium ethanoate, except that one of the hydrogens has been replaced by a methyl group. The overall reaction using two amine molecules is:

$$(CH_3CH_2CO)_2O + 2CH_3NH_2 \rightarrow CH_3CH_2CONHCH_3 + CH_3COONH_3CH_3$$

The mechanism is nucleophilic addition–elimination (Figure 41).

Knowledge check 23

Write an equation for the reaction of propanoyl chloride with methylamine and name the products.

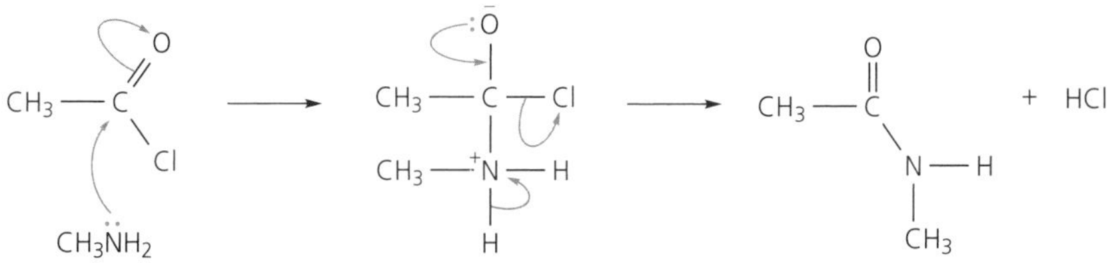

Figure 41 Nucleophilic addition-elimination — reaction of ethanoyl chloride with methylamine

Exam tip

Remember for examination answers the mechanism for nucleophilic addition–elimination needs to show five curly arrows, as shown.

Aspirin is manufactured by acylating 2-hydroxybenzenecarboxylic acid. The –OH group is esterified to form aspirin.

The **industrial advantages** of using ethanoic anhydride to acylate rather than ethanoyl chloride include the following:

- It is less corrosive.
- It is less vulnerable to hydrolysis.
- It is less hazardous to use as it gives a less violent reaction.
- It is cheaper than ethanoyl chloride.
- It does not produce corrosive fumes of hydrogen chloride.

Preparation of a pure organic solid and test of its purity

Organic solids, such as aspirin must be produced in as pure a state as possible. The following method can be used to prepare aspirin, or any organic solid, in the laboratory:

1. Place 20.0 g of 2-hydroxybenzoic acid in a pear-shaped flask and add 40 cm^3 of ethanoic anhydride ($(CH_3CO)_2O$).
2. Safely add 5 cm^3 of conc. phosphoric(v) acid and heat under reflux for 30 minutes.
3. Add water to hydrolyse any unreacted ethanoic anhydride to ethanoic acid, and pour the mixture onto 400 g of crushed ice in a beaker.
4. The product is removed by suction filtration, recrystallised from water and dried in a desiccator.
5. The melting point is then determined.

$$HOOCC_6H_4OH + (CH_3CO)_2O \rightarrow HOOCC_6H_4OCOCH_3 + CH_3COOH$$

Recrystallisation is an important technique used to purify solids by removing unwanted by-products. The method is as follows:

1. Dissolve the impure crystals in the minimum volume of hot solvent.
2. Filter the hot solution by gravity filtration, using a hot funnel and fluted filter paper, to remove any insoluble impurities (filtering through a hot filter funnel and using fluted paper prevents precipitation of the solid).
3. Allow the solution to cool and crystallise (the impurities will remain in solution).
4. Filter off the crystals using suction filtration (Figure 42) (suction filtration is faster than gravity filtration and gives a drier solid).
5. Dry between two sheets of filter paper.

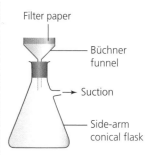

Figure 42 Suction filtration apparatus

A minimum volume of hot solvent is used to dissolve the solid, making a saturated solution. The solution is cooled and the solubility of the compound drops, causing it to recrystallise from solution. Impurities remain dissolved in the solution. A *minimum volume of hot solvent* is used to ensure that as much of the solute is obtained as possible.

To check the **purity** of a solid, a **melting point** can be determined using the following method:

1. Place some of the solid in a melting point tube.
2. Place in melting point apparatus and heat slowly.
3. Record the temperature at which the solid starts to melt and the temperature at which it finishes melting.
4. Repeat and average the temperatures.
5. Compare the melting point with known values in a data book.

Exam tip

The melting point of a substance is the range of temperatures from when the sample starts to melt until it has completely melted. The greater the range the more impurities are present. A range of less than 2°C indicates a fairly pure substance.

Required practical 10

You will be required to prepare a pure sample of an organic solid and test its purity (for example preparation of aspirin).

Summary

- Carboxylic acids are weak acids, but react with carbonates to liberate carbon dioxide.
- Carboxylic acids react with alcohols in the presence of conc. sulfuric acid catalyst to produce esters, which are used in solvents, plasticisers, perfumes and food flavourings.
- Esters can be hydrolysed in acid conditions to produce alcohols and carboxylic acids, and in alkaline conditions to produce alcohols and carboxylic acid salts.
- Vegetable oils and animal fats are esters of propane-1,2,3-triol (glycerol) and can be hydrolysed to produce long-chain carboxylic acid salts (soap) and glycerol.
- Biodiesel is a renewable fuel that consists of a mixture of methyl esters of long-chain carboxylic acids. It is produced by heating vegetable oils with methanol in the presence of an acid catalyst.
- Acyl chlorides (RCOCl), acid anhydrides ($(RCO)_2O$) and amides ($RCONH_2$) all react with water, alcohols, ammonia and primary amines in acylation reactions, where a hydrogen molecule is replaced by an acyl group (RCO–). The mechanism for these reactions is nucleophilic addition–elimination.

Aromatic chemistry

Bonding in benzene

Today's accepted structure for benzene is of a **delocalised model**, which has the following features:

■ It is a **planar hexagonal** molecule of six carbon atoms.

■ All carbon–carbon bond lengths are intermediate in length between that of a single C–C and a double C=C.

■ Each carbon uses three of its outer electrons to form **σ** bonds to two other carbon atoms, and one hydrogen atom. This leaves each carbon atom with one electron in a p orbital. The p orbitals overlap sideways and the six *p* electrons **delocalise** and give regions of electron density above and below the ring (Figure 43).

Exam tip

A circle is used to represent the ring of delocalised electrons:

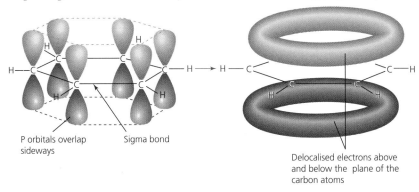

P orbitals overlap sideways

Sigma bond

Delocalised electrons above and below the plane of the carbon atoms

Figure 43 The formation of the delocalised electron structure of benzene

Delocalised electrons Bonding electrons that are not fixed between two atoms, but shared by three or more atoms.

Thermochemical data: enthalpy of hydrogenation

When cyclohexene, with one carbon–carbon double bond, is hydrogenated (Figure 44) the enthalpy of hydrogenation is $-120\,\text{kJ}\,\text{mol}^{-1}$.

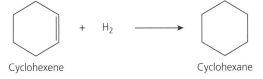

Cyclohexene   Cyclohexane

Figure 44 Hydrogenation of cyclohexene

A theoretical cyclic compound with three double bonds is cyclohexa-1,3,5-triene. It would be expected to have an enthalpy change of $-360\,\text{kJ}\,\text{mol}^{-1}$, because three bonds are being broken (Figure 45).

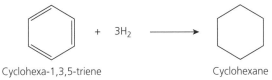

Cyclohexa-1,3,5-triene   Cyclohexane

Figure 45 Hydrogenation of cyclohexa-1,3,5-triene

Enthalpy of hydrogenation is the change in enthalpy that occurs when 1 mol of an unsaturated compound reacts with an excess of hydrogen to become fully saturated.

However, when benzene is hydrogenated the enthalpy change is only $-208\,\text{kJ}\,\text{mol}^{-1}$, which is $152\,\text{kJ}\,\text{mol}^{-1}$ less than expected. This is due to delocalisation of *p* electrons, which makes benzene *more stable* than the theoretical molecule cyclohexa-1,3,5-triene. The delocalised electrons are more spread out and so have fewer electron–electron repulsions.

Electrophilic substitution reactions

Benzene does not take part in addition reactions because the electrons from the delocalised system would need to bond to the atom or groups being added, disrupting the delocalised ring and this would result in a less stable product. Instead benzene undergoes substitution reactions, where one or more of the hydrogen atoms is replaced by another atom or group. The organic product formed retains the delocalised ring of electrons and hence the stability of the benzene ring.

The region of high electron density above and below the plane of the molecule results in benzene being attacked by electrophiles. The mechanism is described as electrophilic substitution.

Nitration

In the nitration of benzene a nitro group (NO_2) replaces one of the hydrogen atoms (Figure 46).

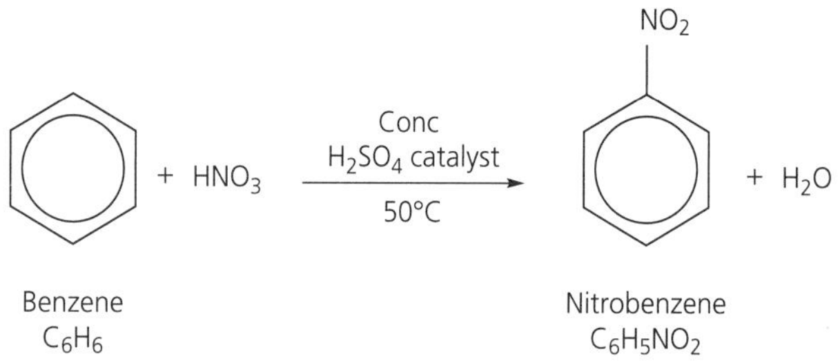

Benzene
C_6H_6

Nitrobenzene
$C_6H_5NO_2$

Figure 46 Nitration of benzene

Conditions: concentrated sulfuric acid and concentrated nitric acid (nitrating mixture) at 50°C.

The overall equation for the generation of the electrophile NO_2^+ (nitronium ion) is:

$$HNO_3 + 2H_2SO_4 \rightarrow NO_2^+ + 2HSO_4^- + H_3O^+$$

The mechanism for mononitration is shown in Figure 47.

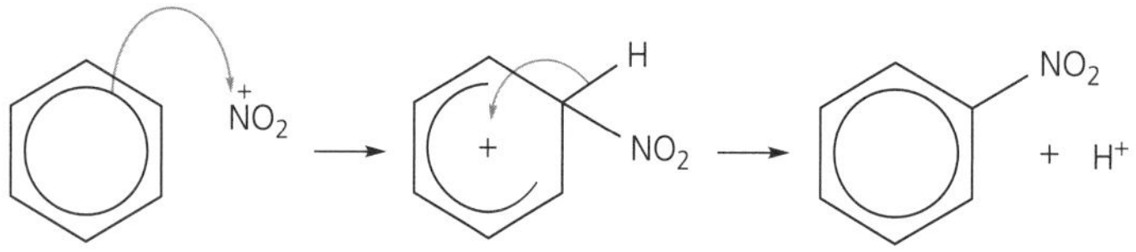

Figure 47 The electrophilic substitution mechanism for the nitration of benzene

The concentrated sulfuric acid acts as a catalyst in the reaction because it is regenerated in the last step when H^+ ion is released in the mechanism and combines with HSO_4^- to reform sulfuric acid.

Knowledge check 24

What is the total number of electrons involved in bonding in benzene?

An **electrophile** is an electron pair acceptor.

Substitution is where one atom or group is replaced by a different atom or group.

Exam tip

In the intermediate the broken electron ring must not extend beyond carbon 2 or carbon 6, though it may be shorter than this.

Nitrobenzene is not prepared in the laboratory due to the high toxicity of benzene. To illustrate nitration in the laboratory, methyl benzoate, a compound of low toxicity is used. The nitrating mixture is added slowly to the methyl benzoate, keeping the temperature less than 10°C. The reaction mixture is poured onto crushed ice and crystalline methyl 3-nitrobenzoate is formed, which is filtered off using suction filtration, washed with cold water, recrystallised and dried. The melting point of the methyl 3-nitrobenzoate is then determined.

Uses of nitration

In organic chemistry the synthesis of a particular compound is often a multistage process in which nitration is one of the steps. Nitration is an important step in the manufacture of explosives such as TNT and in the formation of amines such as phenylamine to produce dyes.

Acylation

Benzene can be acylated using an acyl chloride, in the presence of a catalyst, to form an aromatic ketone (Figure 48). This is an electrophilic substitution reaction in which an acyl group is attached to the ring. Acylation reactions are often referred to as Friedel–Crafts reactions, after the chemists who developed them.

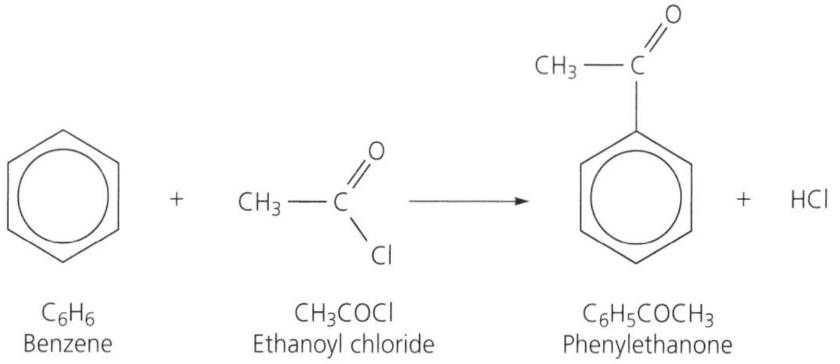

C_6H_6	CH_3COCl	$C_6H_5COCH_3$
Benzene	Ethanoyl chloride	Phenylethanone

Figure 48 Acylation of benzene

Conditions: catalyst of aluminium chloride; anhydrous conditions to prevent hydrolysis of the catalyst.

The required electrophile is the **acylium** ion:

$$CH_3 - \overset{+}{C} = O$$

It is formed by reaction between the ethanoyl chloride and the aluminium chloride catalyst. The equation for the formation of the electrophile is:

$$CH_3COCl + AlCl_3 \rightarrow CH_3\overset{+}{C}O + AlCl_4^-$$

The mechanism for acylation is shown in Figure 49.

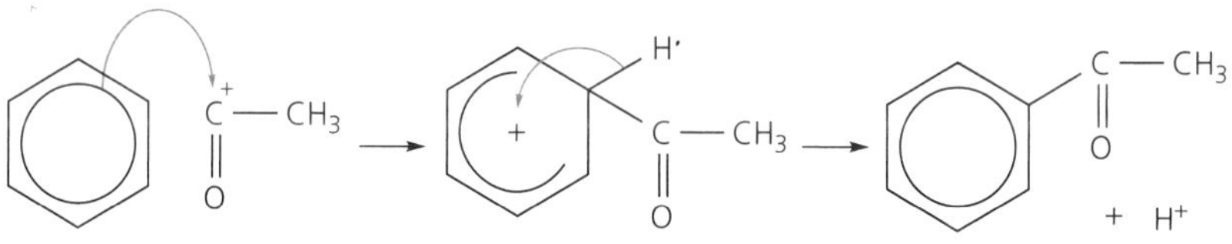

Figure 49 Electrophilic substitution mechanism for the acylation of benzene

Exam tip

The -one in the name phenylethanone shows that it is a ketone and has a C=O in the chain. The phenyl indicates that a C_6H_5 is attached.

The catalyst is regenerated:

$$H^+ + AlCl_4^- \rightarrow AlCl_3 + HCl$$

When methyl benzene reacts with ethanoyl chloride in the presence of aluminium chloride the reaction is faster than the reaction of benzene. This is because the methyl group increases the electron density on the benzene ring, which means the electrophile is attracted more.

Importance of Friedel–Crafts acylation

It is useful in synthesis as the benzene forms a bond with a carbon, producing a side chain. Acylation introduces a reactive carbonyl functional group to the ring, which can undergo the reactions of a carbonyl group and so act as an intermediate in the synthesis of other compounds.

Summary

- Nitration is an electrophilic substitution reaction of benzene in which the nitronium ion is the electrophile generated by the reaction between the reagents concentrated sulfuric and concentrated nitric acid.

- Nitration is important in the manufacture of explosives and the formation of amines.
- Acylation is an electrophilic substitution reaction of benzene using aluminium chloride as a catalyst. It is important in synthesis.

Amines

The structure of amines

Amines are compounds based on ammonia — the hydrogen atoms are replaced with alkyl or aryl (C_6H_5-) groups. Amines contain the $-NH_2$ functional group, which is called the **amino** group.

- A **primary amine** contains *one* alkyl or aryl group attached to the nitrogen atom as only one hydrogen atom in ammonia has been replaced. This means there is only one carbon chain attached to the nitrogen.
- A **secondary amine** contains *two* alkyl or aryl groups attached to the nitrogen atom as two hydrogen atoms in ammonia have been replaced.
- A **tertiary amine** contains *three* alkyl or aryl groups attached to the nitrogen atom as three hydrogen atoms in ammonia have been replaced.

Quaternary ammonium compounds are produced from tertiary amines when the nitrogen's lone pair of electrons forms a dative covalent bond to a fourth alkyl group. Hence a quaternary ammonium salt has four alkyl groups attached to the nitrogen atom. Like ammonium salts they are crystalline ionic solids. For example, $(CH_3)_4N^+I^-$ is tetramethylammonium iodide, a quaternary salt (Figure 50).

Exam tip

Remember that the nitrogen in ammonia can form a dative covalent bond with another hydrogen, producing a quaternary ammonium salt.

$$CH_3 - \overset{\displaystyle CH_3}{\underset{\displaystyle CH_3}{\overset{|}{\underset{|}{N^+}}}} - CH_3 \quad I^-$$

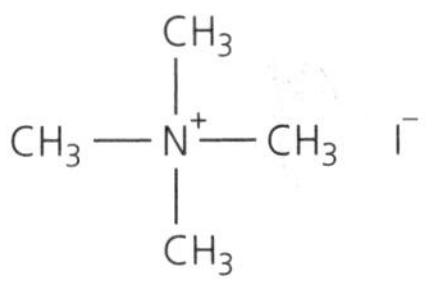

Figure 50 Tetramethylammonium iodide

Exam tip

The easiest way to recognise a primary, secondary or tertiary amine is to count the number of H atoms on the N atom:

- **two** H atoms on N atom — primary amine, for example $C_2H_5N\mathbf{H_2}$
- **one** H atom on N atom — secondary amine, for example $(C_2H_5)_2N\mathbf{H}$
- **no** H atoms on N atom — tertiary amine, for example $(C_2H_5)_3N$

When naming amines there are many variations on the names. Table 9 shows some examples.

Table 9 Names of amines

Structure	Name
$CH_3CH_2CH_2NH_2$	Propylamine (this may also be called aminopropane)
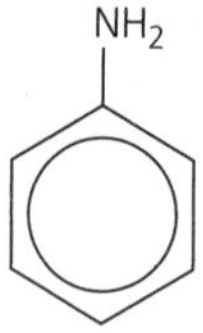 (benzene ring with NH_2)	Phenylamine
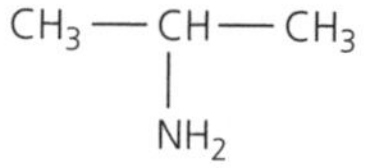 $CH_3 — CH — CH_3$ with NH_2 below	2-aminopropane
$H_2NCH_2CH_2NH_2$	1,2-diaminoethane The prefix **diamino** is used if a compound contains two amino groups
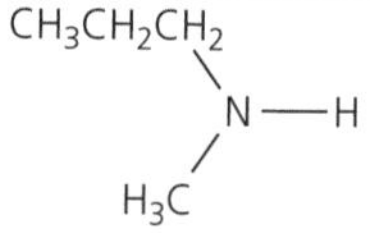 $CH_3CH_2CH_2$ and H_3C attached to $N — H$	N-methylpropylamine The prefix N is used to show that the alkyl groups are attached to the main chain via the nitrogen atom
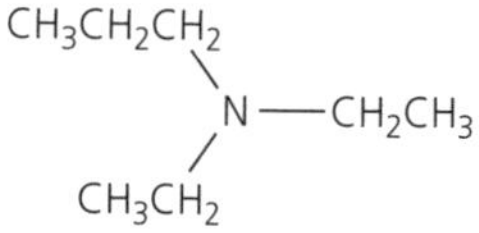 $CH_3CH_2CH_2$ and CH_3CH_2 attached to $N — CH_2CH_3$	N,N-diethylpropylamine

Exam tip

When naming amines, if you need to give the position of the carbon to which the NH_2 group is attached, use the 'amino' form of naming.

Knowledge check 25

Name the compound $(CH_3)_2NH$:

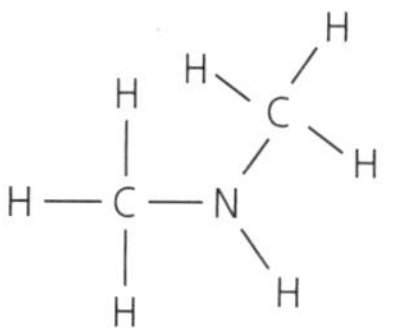

Preparation of primary amines

From halogenoalkanes

This reaction happens in two stages. Initially a salt is formed:

$$CH_3CH_2Br + NH_3 \rightarrow \quad CH_3CH_2NH_3Br$$
$$\text{ethylammonium bromide}$$

Then the salt reacts with the excess ammonia in the mixture and removes a hydrogen ion forming a primary amine:

$$CH_3CH_2NH_3Br + NH_3 \rightarrow CH_3CH_2NH_2 + NH_4Br$$
$$\text{ethylamine}$$
$$\text{(primary amine)}$$

Overall this can be represented by:

$$R\text{–}X \quad + \quad 2NH_3 \quad \rightarrow \quad R\text{–}NH_2 \quad + \quad NH_4X$$
$$\text{halogenoalkane} \quad \text{ammonia} \quad \text{amine} \quad \text{ammonium halide}$$

For example:

$$CH_3CH_2Cl \quad + \quad 2NH_3 \quad \rightarrow \quad CH_3CH_2NH_2 \quad + \quad NH_4Cl$$
$$\text{chloroethane} \quad \text{ammonia} \quad \text{ethylamine} \quad \text{ammonium chloride}$$

Conditions: heat in a sealed flask with excess ammonia in ethanol. A sealed glass tube is used because the ammonia would escape as a gas if reflux was implemented.

From nitriles by reduction

Amines can be formed by the reduction of nitriles using hydrogen in the presence of a nickel catalyst or by using a reducing agent such as lithium tetrahydridoaluminate(III) (LiAlH$_4$) in ether:

$$R-C≡N \ + \ 2H_2 \ \rightarrow \ R-CH_2NH_2$$

For example:

$$\underset{\text{ethanenitrile}}{CH_3CN} \ + \ 2H_2 \ \rightarrow \ \underset{\text{ethylamine}}{CH_3CH_2NH_2}$$

Conditions: hydrogen in the presence of a nickel catalyst.

$$R-C≡N \ + \ 4[H] \ \rightarrow \ R-CH_2NH_2$$

For example:

$$\underset{\text{propanenitrile}}{CH_3CH_2CN} \ + \ 4[H] \ \rightarrow \ \underset{\text{propylamine}}{CH_3CH_2CH_2NH_2}$$

Conditions: [H] is lithium aluminium hydride in dry ether.

Preparation of aromatic amines

Aromatic amines are prepared by reduction of nitrocompounds — for example, phenylamine is prepared by reduction of nitrobenzene using tin and concentrated hydrochloric acid as a reducing agent (Figure 51):

$$RNO_2 + 6[H] \rightarrow RNH_2 + 2H_2O$$

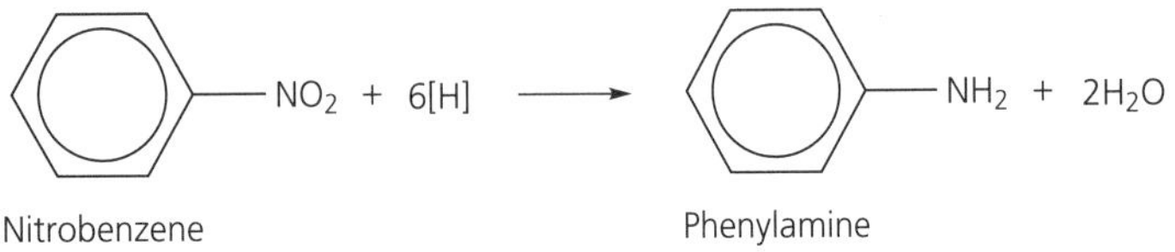

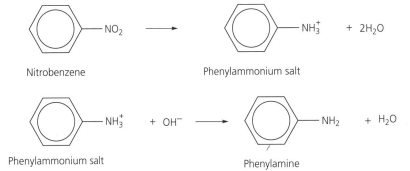

Figure 51

Conditions: heat under reflux with tin and excess concentrated hydrochloric acid, followed by adding concentrated sodium hydroxide.

Because excess acid is used, rather than getting phenylamine directly, the protonated form of phenylamine — **phenylammonium ion** — is formed and the concentrated sodium hydroxide removes the hydrogen ion, liberating the free amine (Figure 52).

Figure 52

> **Exam tip**
>
> The reduction of a nitrile results in an increase in the length of the carbon chain. This is useful in organic synthesis.

> **Exam tip**
>
> The reduction reaction gives a better yield than the preparation of amines from halogenoalkanes, and there are no other products.

> **Exam tip**
>
> If a compound containing two nitro groups is reduced, both nitro groups will change to NH$_2$ groups.

Aromatic amines prepared by the reduction of nitro compounds are used in the manufacture of dyes.

Basic properties of amines

A base is a proton acceptor. Amines are weak bases because the lone pair of electrons on the nitrogen atom can accept a proton, as shown in Figure 53.

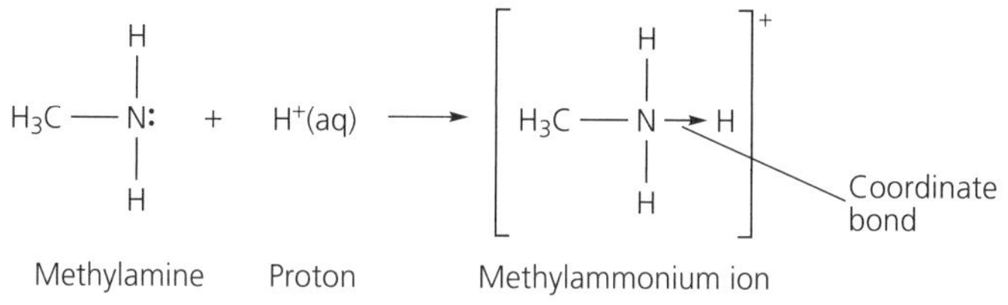

Methylamine   Proton   Methylammonium ion

Figure 53

Basic reactions of amines

With dilute mineral acids

amine + acid → alkyl ammonium salt

With dilute hydrochloric acid (Figure 54):

$$CH_3NH_2 \quad + \quad HCl \quad \rightleftharpoons \quad CH_3NH_3{}^+Cl^-$$
methylamine   methylammonium chloride

Phenylamine   Phenylammonium chloride

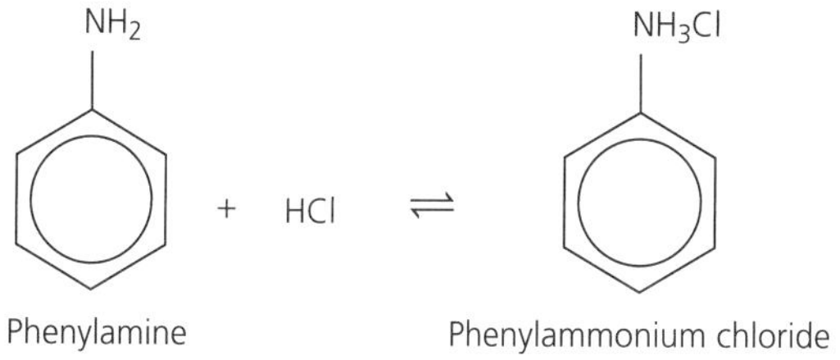

Figure 54

With dilute sulfuric acid:

$$2CH_3CH_2NH_2 \quad + \quad H_2SO_4 \rightleftharpoons \quad (CH_3CH_2NH_3)_2SO_4$$
ethylamine   ethylammonium sulfate

With water

When amines react with water, they accept a hydrogen ion from water to produce an alkylammonium ion and hydroxide ions.

$$CH_3NH_2 \quad + \quad H_2O \rightleftharpoons \quad CH_3NH_3{}^+ \quad + \quad OH^-$$
methylamine   methylammonium ion

The solution formed is weakly basic because the equilibrium lies to the left as methylamine is only partly ionised. As a result, little of it has reacted with the water, resulting in a solution with low [OH⁻].

Comparison of base strength

Different types of amine have different basic strengths.

- **Primary aliphatic amines** are stronger bases than ammonia because of the alkyl group attached to the nitrogen. The alkyl group is said to be **electron donating** — it releases electrons, meaning there is slightly more electron density on the nitrogen atom. As a result, the *lone pair is more available* and so has an increased ability to accept a proton. Aliphatic amines generally increase in base strength as the number of alkyl groups attached to the nitrogen atom increases.
- **Primary aromatic amines** are weaker bases than ammonia because nitrogen's lone pair of electrons can overlap with the delocalised π electrons in the benzene ring. The lone pair is delocalised into the π system, the electron density on the nitrogen is decreased and the *lone pair is less available* for accepting a proton.

Knowledge check 26

Place ammonia, propylamine and phenylamine in order of increasing basic strength.

Nucleophilic properties of amines

All amines contain a lone pair of electrons on the nitrogen atom, so they act as nucleophiles.

Nucleophilic substitution reaction with halogenoalkanes

The reaction of halogenoalkanes with ammonia and amines forms primary, secondary, tertiary amines and quaternary ammonium salts. The reaction to prepare a primary amine has been mentioned on page 51.

Making a secondary amine

Further substitution reactions can occur. This is because the primary amine produced has a lone pair, so it can act as a nucleophile and continue to react with any unused halogenoalkane, in the same stages as before:

$$CH_3CH_2Br + CH_3CH_2NH_2 \rightarrow (CH_3CH_2)_2NH_2Br$$
$$\text{diethylammonium bromide}$$

$$(CH_3CH_2)_2NH_2Br + NH_3 \rightleftharpoons (CH_3CH_2)_2NH + NH_4Br$$
$$\text{diethylamine}$$
$$\text{(secondary amine)}$$

This successive substitution results in the formation of a secondary amine.

Making a tertiary amine

The secondary amine produced also has a lone pair so it too can act as a nucleophile and continue to react with any unused halogenoalkane, in the same stages as before. This further substitution results in a tertiary amine:

$$CH_3CH_2Br + (CH_3CH_2)_2NH \rightarrow (CH_3CH_2)_3NHBr$$

$$(CH_3CH_2)_3NHBr + NH_3 \rightleftharpoons (CH_3CH_2)_3N + NH_4Br$$
$$\text{triethylamine}$$
$$\text{(tertiary amine)}$$

Making a quaternary ammonium salt

Finally, the tertiary amine reacts with the halogenoalkane to form a quaternary ammonium salt. There is no longer a lone pair on the nitrogen so the quaternary ammonium salt cannot act as a nucleophile and the reaction stops:

$$CH_3CH_2Br + (CH_3CH_2)_3N \rightarrow (CH_3CH_2)_4NBr$$
$$\text{quaternary ammonium salt}$$

Choosing the conditions

The initial conditions can be adjusted to favour production of a particular type of amine.

- **Excess ammonia** favours the production of **primary amines** because it is less likely that another halogenoalkane molecule will react with an amine when there is a large number of unreacted ammonia molecules available.
- **Excess halogenoalkane** favours the production of the **quaternary ammonium salts** because it ensures that each ammonia reacts with four halogenoalkane molecules.

The use of quaternary ammonium salts

Quaternary ammonium salts are used in the production of **cationic surfactants**, which are found in detergents, fabric softeners and hair conditioners. They coat the surface of the cloth or hair with positive charges and reduce the static due to negatively charged electrons.

Mechanism

The reactions of a halogenoalkane with ammonia and amines, forming primary, secondary and tertiary amines, and quaternary ammonium salts, are nucleophilic substitution reactions. The amines have lone pairs and are nucleophiles (Figure 55).

$$CH_3CH_2Br + 2NH_3 \rightarrow CH_3CH_2NH_2 + NH_4Br$$
$$\text{bromoethane} \qquad\qquad \text{ethylamine} \qquad \text{ammonium bromide}$$

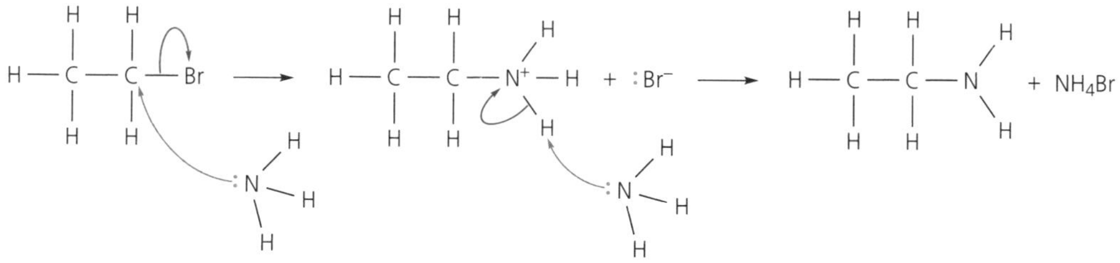

Figure 55

Further substitution is possible as the product, the primary amine, is also a nucleophile (Figure 56).

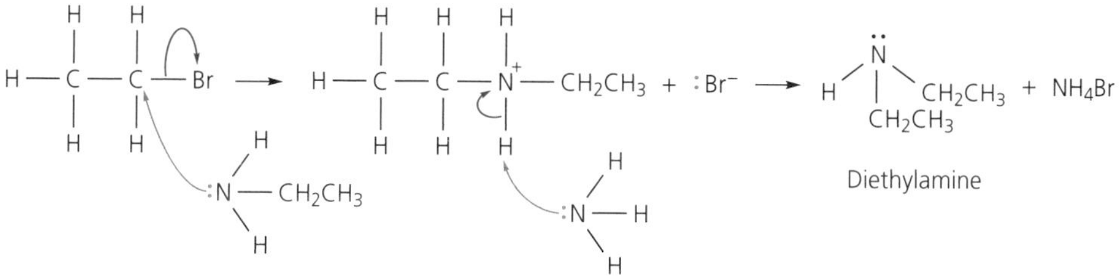

Diethylamine

Figure 56

The mechanism for the formation of a quaternary ammonium ion is shown in Figure 57.

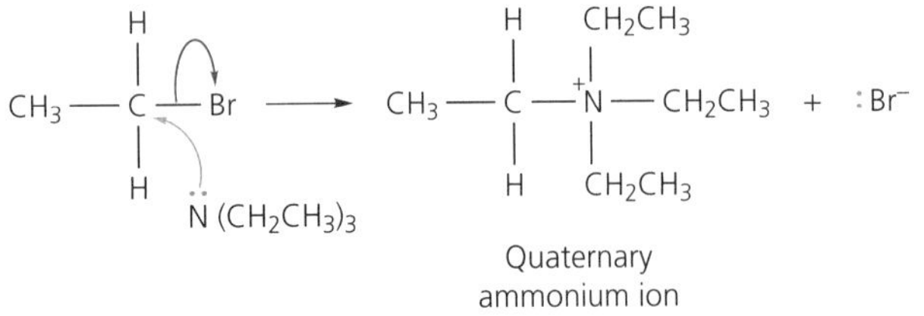

Figure 57

Exam tip

The nucleophilic addition–elimination reactions of ammonia and primary amines with acyl chlorides are covered on pages 43–45.

Summary

- Amines, like ammonia, are weak bases and react with acids.
- Primary aliphatic amines are stronger bases than ammonia because of the electron-donating alkyl group making the lone pair more available.
- Primary aromatic bases are weaker bases than ammonia because nitrogen's lone pair is delocalised into the π system and is less available for accepting a proton.
- Aromatic amines, used in dye production, are prepared by reduction of nitro compounds using tin and concentrated hydrochloric acid as a reducing agent.
- Primary aliphatic amines are prepared by the reduction of nitriles using hydrogen in the presence of a nickel catalyst, or $LiAlH_4$ as a reducing agent.
- Primary amines are prepared in the nucleophilic substitution reaction of excess ammonia with halogenoalkanes; using excess halogenalkane will produce secondary and tertiary amines.
- A quaternary ammonium salt is formed from tertiary amines when the nitrogen's lone pair forms a dative covalent bond with a fourth alkyl group. Such salts are cationic surfactants.

Polymers

Polymers are synthetic and natural molecules that have a repeating structure. There are several types including common plastics, proteins and DNA.

Exam tip

Addition polymers such as polythene and PVC were discussed in the first year of this course. Make sure you read over this section to familiarise yourself with it again.

A **polymer** is a long chain of repeating monomer units joined together.

Condensation polymers

A condensation polymer is a polymer formed when two monomers react, releasing a small molecule such as H_2O or HCl. There are two main types of condensation polymer: polyesters and polyamides.

Polyesters

A polyester is made when many ester groups are formed between monomers, creating a long-chain molecule. The most common example is polyethylene terephthalate. It is the polymer formed from ethane-1,2-diol and terephthalic acid (benzene-1,4-dicarboxylic acid). It is often abbreviated to PET or PETE for recycling purposes. Almost all water bottles are made from PET. Figure 58 shows its formation.

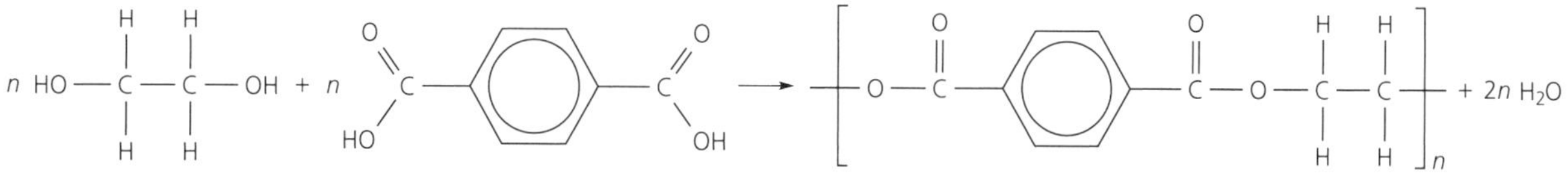

Ethane-1,2-diol | Benzene-1,4-dicarboxylic acid (terephthalic acid) | Poly(ethylene terephthalate)

Figure 58 Formation of PET

Exam tip

The structure shown in brackets is the repeating structure of the polymer. A question may ask you to show the repeating unit of the polymer. Make sure you can draw the structure of the two monomers for this polymer and be able to recognise them to identify the polymer they would form.

The diacyl chloride of benzene-1,4-dicarboxylic acid could be used and 2HCl would be eliminated instead of $2H_2O$.

PET is a biodegradable polymer because ester groups can be hydrolysed, so it is environmentally preferable to addition polymers, which are not hydrolysed. It is always better to recycle polymers, including biodegradable ones such as PET.

Knowledge check 27

State the IUPAC names of the monomers used to form PET.

Polyamides

A polyamide is made when many amide groups are formed between molecules, creating a long chain molecule. The most common example is nylon-6,6. This is the polymer formed from hexane-1,6-diamine and hexane-1,6-dioic acid. Figure 59 shows its formation.

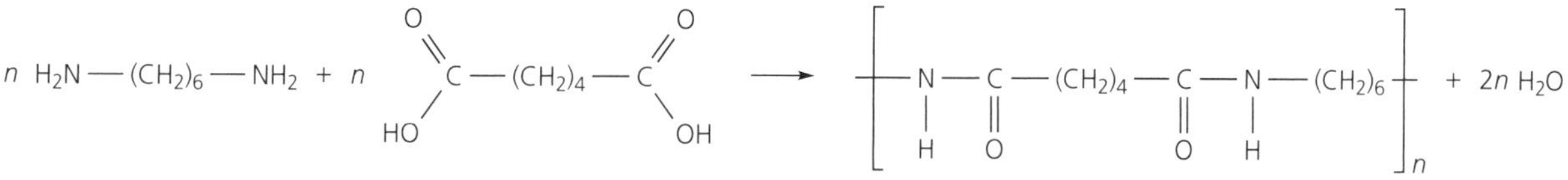

Hexane-1,6,diamine | Hexane-1-6-dioic acid | Nylon

Figure 59

The polyamide is called nylon-6,6 because there are six carbon atoms in each of the monomers. Nylon-6,10 is another nylon, where the acid has 10 carbon atoms. Nylon-6,6 is used to make tights and ropes.

The polymer can also be made using the diacyl chloride and, again, 2HCl are eliminated instead of $2H_2O$ (Figure 60).

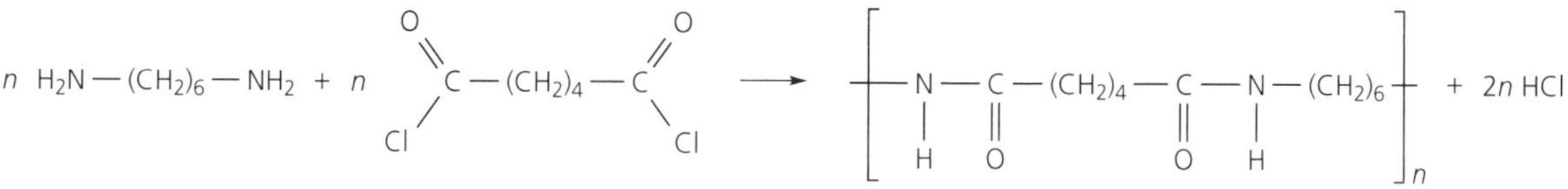

Hexane-1,6,diamine | Hexanedioyl dichloride | Nylon

Figure 60

Another polyamide is Kevlar. Kevlar is used in bullet-proof vests. It is formed from benzene-1,4-dicarboxylic acid and benzene-1,4-diamine. The equation for its formation is shown in Figure 61.

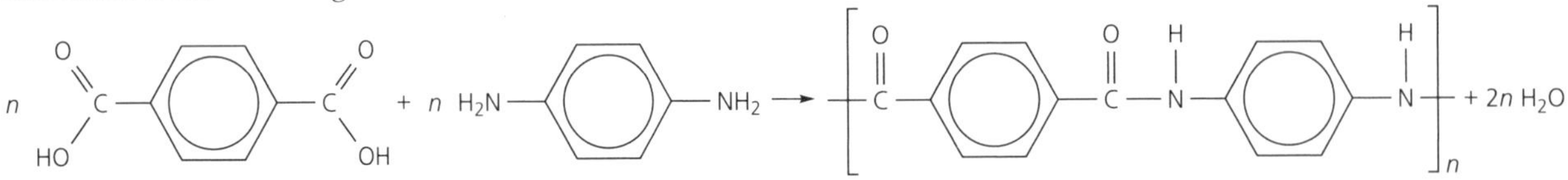

Figure 61 Formation of Kevlar

There are hydrogen bonds between the chains of polyamides and permanent dipole attractions between the chains of polyesters. This means that these condensation polymers can be formed into fibres, which can be woven into clothing. Addition polymers are mostly non-polar, so cannot be used to manufacture clothing.

Polyamides are also biodegradable because the amide bonds can be hydrolysed. Poly(alkene) polymers are not biodegradable because the bonds they contain cannot be hydrolysed.

Biodegradability and disposal of polymers

Many polymers are now recycled. However, disposal of synthetic polymers creates an ongoing environmental problem. Disposal methods include landfill and incineration.

Addition polymers added to a landfill site may be present for many years because they are non-biodegradable. Landfill sites waste land and they can often be an eyesore. However, landfill sites are local to the community so the waste does not need to be transported over large distances.

Incineration releases greenhouse gases into the air and also some toxic gases depending on the polymers being burned. There is a toxic residue left with incineration but the energy released can be harnessed and used to generate electricity. Incineration plants are generally not local, so the waste has to be transported.

> **Knowledge check 28**
>
> Which of the following are biodegradable: polythene, PET, nylon-6,6, Kevlar?

Summary

- Polymers may be classified as addition polymers (formed from alkenes) or condensation polymers.
- Common addition polymers are polythene or PVC.
- Condensation polymers are polyesters or polyamides
- PET is a polyester whereas nylon and Kevlar are polyamides
- Condensation polymerisation involves the elimination of a small molecule such as water or hydrogen chloride

---

Amino acids, proteins and DNA

Amino acids

An amino acid has an amino group (NH_2) and a carboxylic acid group (COOH). The amino group and the acid group are bonded to the same carbon atom. This carbon atom to which the two groups are bonded is called the α carbon atom, so these are referred to as α amino acids. The structure of an amino acid is shown in Figure 62.

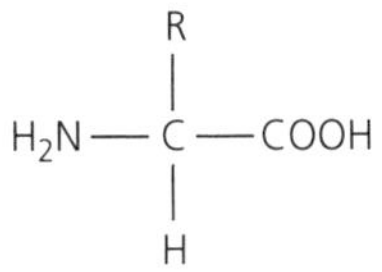

Figure 62

R is a group, which can be H (this amino acid is called glycine), an alkyl group or other groups. There are 20 naturally occurring amino acids. The ones you need to know about are given on your data leaflet for AQA examinations (Figure 63).

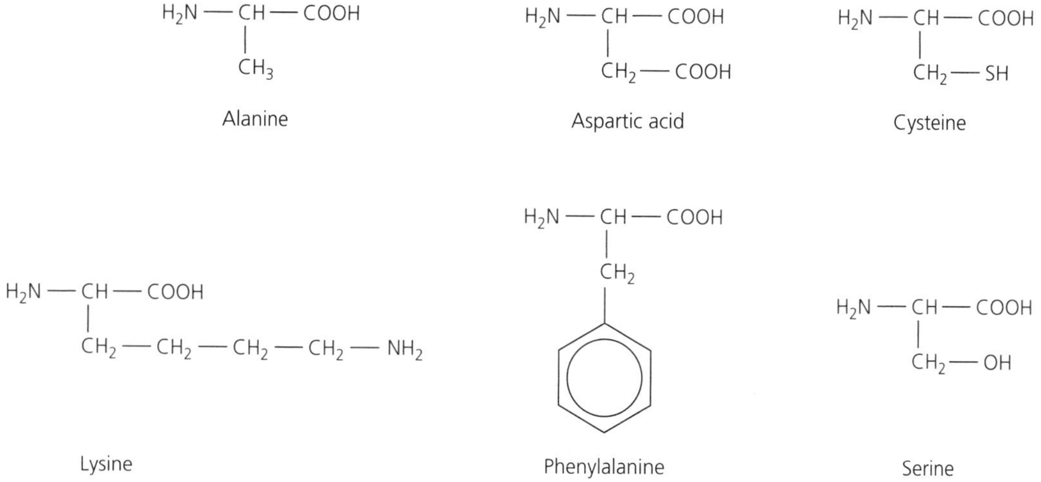

$$H_2N - CH - COOH$$
$$| $$
$$CH_3$$

Alanine

$$H_2N - CH - COOH$$
$$|$$
$$CH_2 - COOH$$

Aspartic acid

$$H_2N - CH - COOH$$
$$|$$
$$CH_2 - SH$$

Cysteine

$$H_2N - CH - COOH$$
$$|$$
$$CH_2 - CH_2 - CH_2 - CH_2 - NH_2$$

Lysine

$$H_2N - CH - COOH$$
$$|$$
$$CH_2$$

Phenylalanine

$$H_2N - CH - COOH$$
$$|$$
$$CH_2 - OH$$

Serine

Figure 63

Amino acids are often categorised based on the side chain. Amino acids with a non-polar R group, such as alanine and phenylalanine, are often compared to those with a polar R group, such as aspartic acid, cysteine, lysine or serine.

Amino acids have unusual properties for organic molecules because they exist as **zwitterions**. A zwitterion is a dipolar ion with both a negative and positive charge. The structure of the dipolar ion is shown in Figure 64.

$$R$$
$$|$$
$$H_3\overset{+}{N} - C - COO^-$$
$$|$$
$$H$$

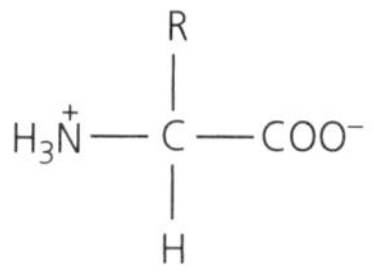

Figure 64

The melting points of amino acids are higher than expected due to the ionic bonding between these dipolar ions. They are soluble in water.

Amino acids in solution

In acidic solution, the NH_3 group is protonated so a positively charge ion is present (Figure 65).

In alkaline solution the COOH group is deprotonated so a negatively charged ion is present (Figure 66).

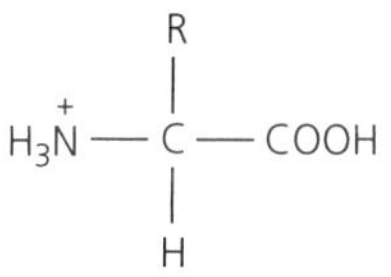

$$R$$
$$|$$
$$H_3\overset{+}{N} - C - COOH$$
$$|$$
$$H$$

Figure 65

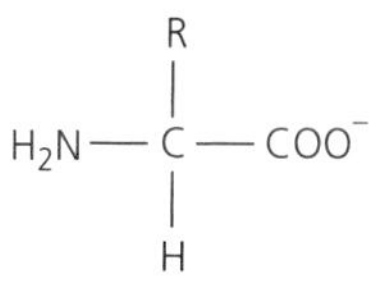

$$R$$
$$|$$
$$H_2N - C - COO^-$$
$$|$$
$$H$$

Figure 66

A **zwitterion** is a dipolar ion with both a positive charge and a negative charge.

Knowledge check 29

Draw the structure of serine in a solution of pH 12.

Exam tip

A question may ask you to show the structure of the ion of the amino acid in solutions with a pH of less than 7 (acidic) or pH greater than 7 (alkaline). Any COOH or NH_2 side groups will also be affected.

Worked example

Show the structure of lysine in a solution of pH 1.

Answer

Lysine has the structure:

H_2N

H — C — CH_2 — CH_2 — CH_2 — CH_2 — NH_2

COOH

In acidic pH, both NH_2 groups will be protonated. The structure is:

$H_3\overset{+}{N}$

H — C — CH_2 — CH_2 — CH_2 — CH_2 — $\overset{+}{N}H_3$

COOH

Nomenclature of amino acids

The IUPAC names of amino acids are often examined.

- Glycine (Figure 67a) is named aminoethanoic acid.
- Alanine (Figure 67b) is named is 2-aminopropanoic acid.
- Lysine (Figure 67c) is named 2,6-diaminohexanoic acid. The acid group has priority so it is named as a carboxylic acid.
- Glutamic acid (Figure 67d) is named is 2-aminopentanedioic acid.

(a)

H_2N

H — C — H

COOH

(b)

H_2N

H — C — CH_3

COOH

(c)

H_2N

H — C — CH_2 — CH_2 — CH_2 — CH_2 — NH_2

COOH

(d)

H_2N

H — C — CH_2 — CH_2 — COOH

COOH

Figure 67

Knowledge check 30

Give the IUPAC name for leucine:

H_3C \ / CH_3
CH
|
CH_2
|
H_2N — C — COOH
|
H

Leucine

Polymerisation of amino acids

Amino acids join together to form condensation polymers. Short-chain polymers are called peptides. A dipeptide has two amino acids. A tripeptide has three amino acids. Polypeptides contain many amino acids in a chain. The amino acids within a peptide are often called amino acid residues. These are joined by peptide links.

Two amino acids react to form a dipeptide, as shown in Figure 68.

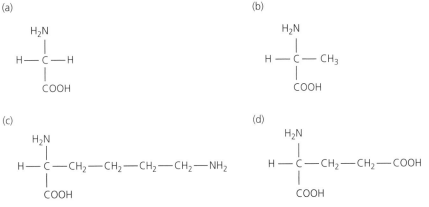

Amino acid 1   Amino acid 2   A dipeptide

Figure 68 Formation of a dipeptide

If the two amino acids are different then there are two different possible dipeptides because the order is important. A question might ask for the peptide link to be circled (Figure 69).

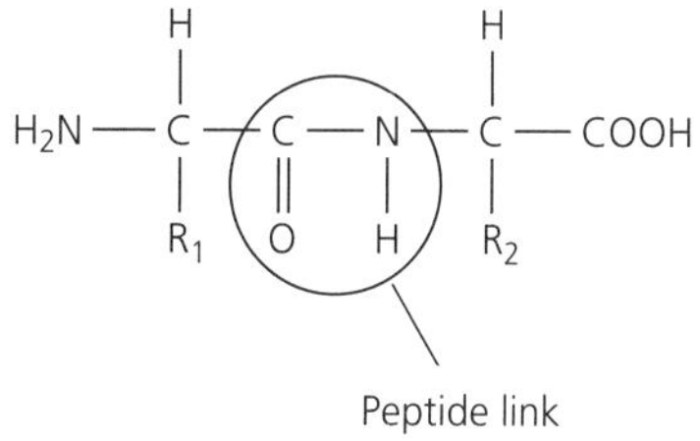

Peptide link

Figure 69 Peptide link

The peptide link is a substituted amide group and is often written as $-CONH-$.

Worked example

Draw the two different dipeptides that can be formed from alanine and phenylalanine. The structures of alanine and phenylalanine (Figure 70) are given in the data leaflet.

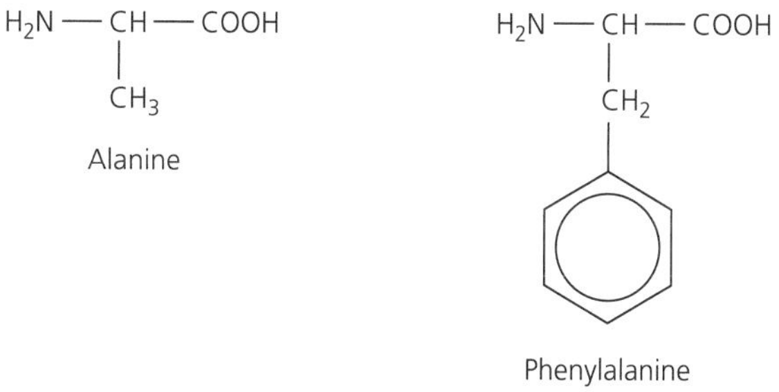

Alanine

Phenylalanine

Figure 70

Answer

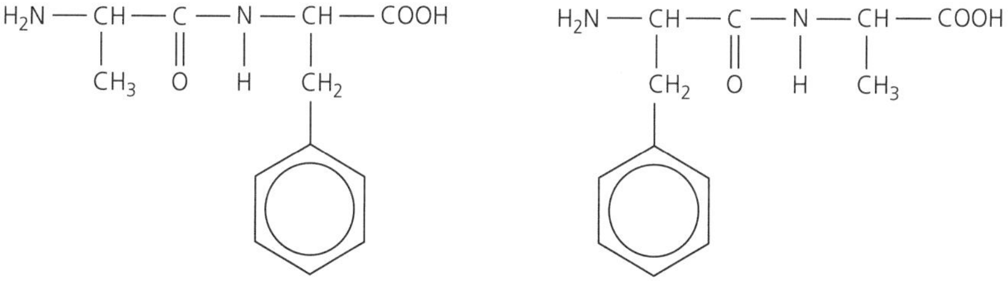

Figure 71

Hydrolysis of a peptide

The peptide link like the amide group can be hydrolysed. Hydrolysis may be in acid or alkaline solution and the structure of the amino acids formed depends on these conditions.

In acid conditions the amino acids formed are protonated, meaning that all NH_2 groups accept a proton. COOH groups stay as COOH.

In alkaline conditions the amino acids formed are deprotonated, meaning that all COOH groups lose a proton to become COO^-. NH_2 groups stay as NH_2.

Exam tip

You could be asked to draw the structure of a peptide containing up to three amino acids.

Worked example

Identify the three amino acids in the tripeptide in Figure 72. Draw the structures of the species obtained on hydrolysis of the tripeptide in acid conditions.

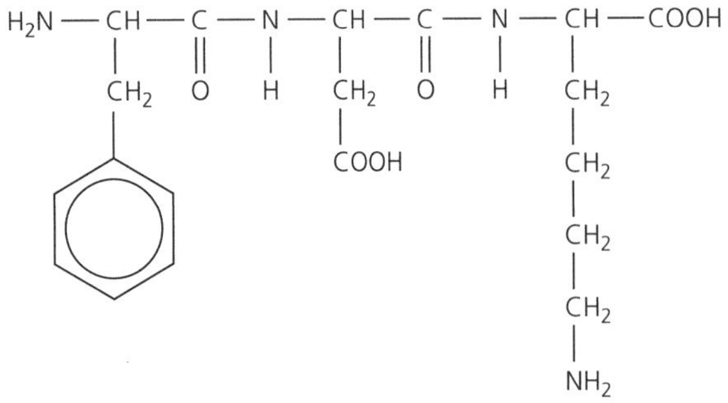

Figure 72

Answer

The three amino acid residues in the tripeptide from left to right are phenylalanine, aspartic acid and lysine.

In acid conditions the amino acids are protonated. The species formed are shown in Figure 73.

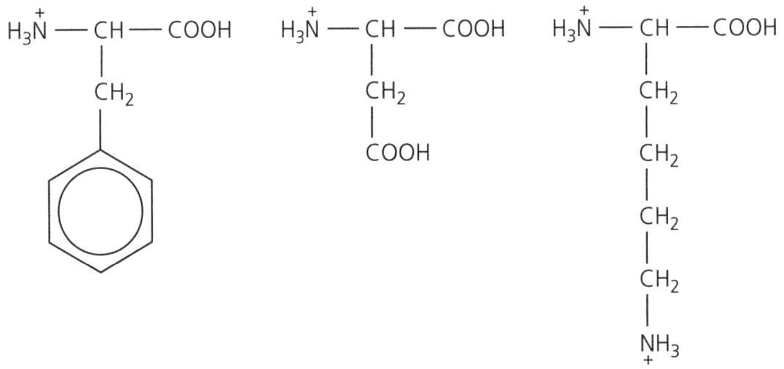

Figure 73

The amino acids that are produced on hydrolysis of a protein can be separated by thin-layer chromatography (TLC). More detail on TLC is given on page 77. Amino acids are colourless, so a chemical stain called ninhydrin is used to locate the position of the spot; or they can be visualised under ultraviolet light. The R_f value of the spots can be used to identify the amino acids present.

Proteins

Proteins are naturally occurring polymers. The structure of a protein may be divided into primary, secondary and tertiary.

The **primary structure** of a protein is the chain of amino acids joined by peptide links.

Exam tip

Use your data leaflet to check that you can identify the amino acids.

Knowledge check 31

State two ways in which amino acids can be viewed on a TLC chromatogram.

There are two main types of **secondary structure** of a protein: α-helix and β-pleated sheet. The chain of amino acids twists into an α-helix and side-by-side chains form β-pleated sheets. The structures are held together by hydrogen bonds between the polar C=O and N–H groups in different peptide groups.

Knowledge check 32

State two features of the secondary structure of a protein.

The **tertiary structure** of a protein is the final folding of the protein chain into its 3D shape. There are many interactions that hold the tertiary structure together, for example:

- hydrogen bonding between polar side-chain groups
- disulfide bridges (–S–S–) between cysteine residues in different parts of the protein chain
- salt bridges between –COO⁻ and –NH₃⁺ groups
- hydrophobic and hydrophilic interactions where hydrophobic side chains fold inside the protein molecule and hydrophilic ones interact with the aqueous environment at the surface of the protein
- minimising steric hindrance between large groups

The structures are summarised in Figure 74.

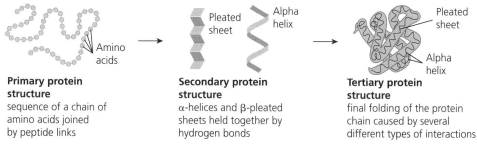

Primary protein structure
sequence of a chain of amino acids joined by peptide links

Secondary protein structure
α-helices and β-pleated sheets held together by hydrogen bonds

Tertiary protein structure
final folding of the protein chain caused by several different types of interactions

Figure 74 The primary, secondary and tertiary structure of a protein

Enzymes

- **Enzymes** are proteins.
- They are biological catalysts and function via an active site on the surface of the molecule.
- The reactant (called a substrate) fits into the active site.
- Enzymes are stereospecific, so only one optical isomer will fit into the active site. The other enantiomer will not fit into the active site because the 3D arrangement of the atoms is not the same.
- There are groups in the active site that catalyse the reaction.
- The enzyme and substrate bind together to form the enzyme–substrate complex.
- Computer modelling of an enzyme can allow for the design of drugs that would inhibit the enzyme by binding to the active site and preventing the enzyme from binding to the substrate.
- Enzymes are affected by changes in temperature and pH. At high temperatures enzymes are denatured (the tertiary structure is disrupted) and enzyme activity decreases.
- Enzymes in the human body operate most effectively in the temperature range 35–40°C.
- Extremes of pH also affect the function of enzymes, though some enzymes like pepsin are designed to work in the stomach in very acidic pH.

Enzymes are biological catalysts.

Content Guidance

DNA

DNA stands for deoxyribonucleic acid. It consists of a sugar–phosphate backbone with bases bonded to the sugar molecules. Two strands of DNA hydrogen bond to each other to form a double helix. The hydrogen bonds are between the bases.

The sugar in DNA is 2-deoxyribose. The structure of ribose and 2-deoxyribose are shown in Figure 75. The carbon atoms are not shown, as in a skeletal formula.

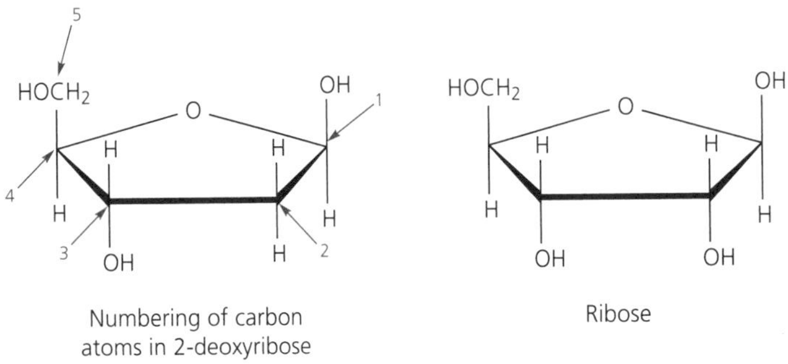

Numbering of carbon atoms in 2-deoxyribose

Ribose

Figure 75 Ribose and 2-deoxyribose

It is important to understand the numbering of the carbon atoms in the sugar so that it is clear why it is 2-deoxyribose. Also, the phosphate and bases are bonded to specific carbon atoms of 2-deoxyribose.

The phosphate group is shown in Figure 76.

The phosphate group links carbon 5 on one 2-deoxyribose molecule to carbon 3 on another (Figure 77).

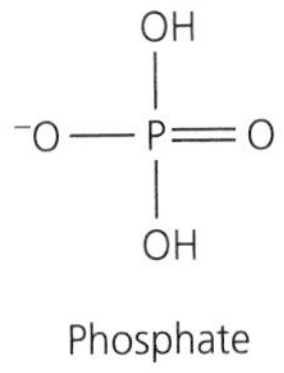

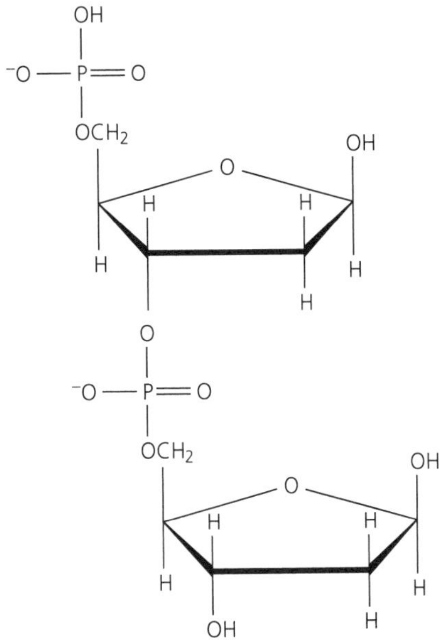

Figure 77 The sugar–phosphate backbone

Phosphate

Figure 76

This sugar–phosphate backbone then has bases bonded to carbon 1 of the 2-deoxyribose. The structures of the four bases in DNA are shown in Figure 78. They are adenine, guanine, cytosine and thymine. Adenine and guanine are purine bases and cytosine and thymine are pyrimidine bases. They all contain an amino group. The nitrogen atom that bonds to carbon 1 in 2-deoxyribose is circled in green.

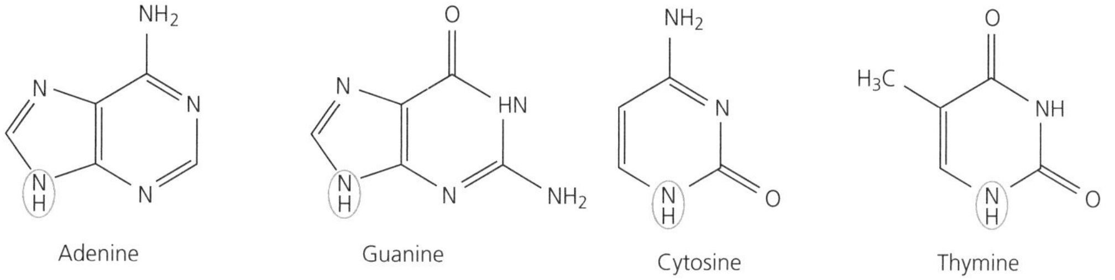

Figure 78 The four bases found in DNA

DNA is synthesised from nucleotides, which consist of a 2-deoxyribose with a phosphate group bonded to carbon 5 and a base bonded to carbon 1. The four nucleotides are shown in Figure 79.

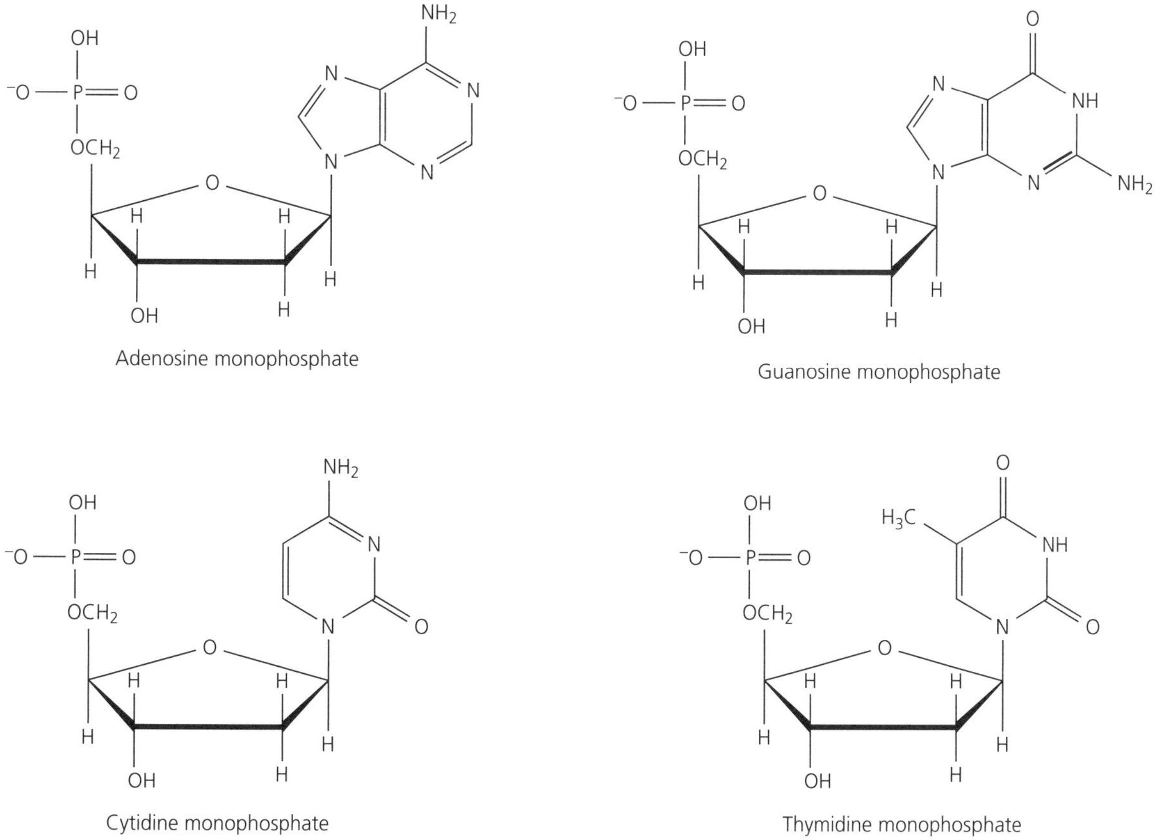

Adenosine monophosphate

Guanosine monophosphate

Cytidine monophosphate

Thymidine monophosphate

Figure 79 The four nucleotides found in DNA

The nucleotides bond together to form a chain of a single strand of DNA. The reaction between the carbon 3 OH group and the OH group of the phosphate bonded to carbon 5 of another nucleotide is a condensation reaction. Water is released.

Another strand of DNA runs in the opposite direction and the two strands pair up. Hydrogen bonds between the bases hold the double-stranded molecule together. Hydrogen bonds form between thymine and adenine and between cytosine and guanine (Figure 80).

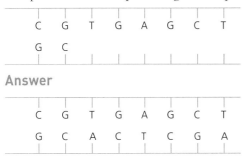

Figure 80 Bonding between bases in DNA

There are two hydrogen bonds between thymine and adenine but there are three hydrogen bonds between cytosine and guanine. For regions that are rich in GC base pairs, this affects the physical properties of DNA. A section of DNA showing the complementary strands is shown in Figure 81. One strand runs from the carbon 5 end (represented as 5') to the carbon 3 end (represented as 3'). The other strand runs from 3' to 5'. The strands are described as complementary because the bases complement each other, for example T on one strand always pairs with A on the other strand. Similarly, G on one strand always pairs with C on the other strand.

DNA structure can be shown simply as a series of bases using the letters G, C, T and A. Some bases may be omitted and you might be asked to complete the sequence.

Worked example

Complete the letters representing the complementary bases in DNA.

C	G	T	G	A	G	C	T
G	C						

Answer

C	G	T	G	A	G	C	T
G	C	A	C	T	C	G	A

Exam tip

The bases are often represented by the letters G for guanine, C for cytosine, T for thymine and A for adenine. It is important to remember that G pairs with C and T pairs with A.

Knowledge check 34

Name the four bases in DNA.

Exam tip

Remember that the base pairs are GC and AT.

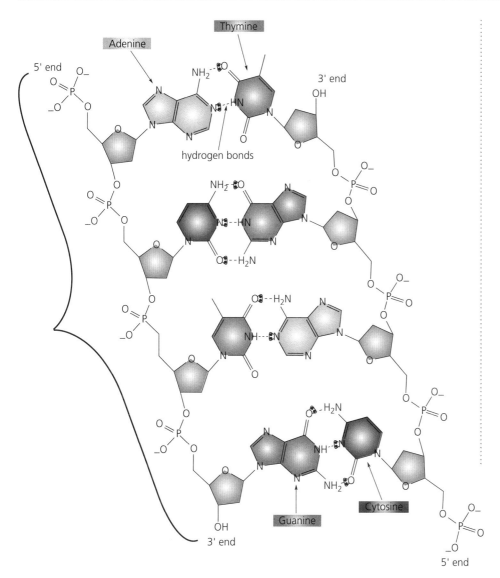

Figure 81

Action of anticancer drugs

Cancer occurs when cell division is unchecked and one cell continues to divide to form a tumour. Many anticancer drugs kill dividing cells.

Cisplatin is a commonly used anticancer drug. It is the *cis* or Z form of $[Pt(NH_3)_2Cl_2]$ and has a square planar shape — its structure was discussed on page 16.

Cisplatin reacts with water. One of the chloro ligands is substituted for a water ligand, as shown in the following equation:

$$[Pt(NH_3)_2Cl_2] + H_2O \rightarrow [Pt(NH_3)_2Cl(H_2O)]^+ + Cl^-$$

The complex $[Pt(NH_3)_2Cl(H_2O)]^+$ then acts as an anticancer drug in two ways:

- A nitrogen atom in guanine bonds to Pt, displacing the water ligand (Figure 82).
- A hydrogen atom on one of the ammonia ligands of cisplatin hydrogen bonds to a nitrogen or oxygen atom in guanine.

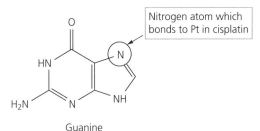

Figure 82

Cisplatin attached to guanine residues in DNA prevents the DNA from being replicated as the enzymes cannot copy this base. The cell then enters programmed cell death and is destroyed.

The risks with anticancer drugs are that they can damage normal cells or may kill the patient. These risks are lowered by giving the drug in short bursts, allowing patients time between treatments and also monitoring them. If possible the drug is targeted at the site of the cancer. Patients undergoing treatment with drugs like cisplatin will often lose their hair and they may no longer be able to reproduce because sex cells are also killed. Medical professionals will advise the patient, but ultimately it is the patient who decides whether to take the treatment or not.

Knowledge check 35

Write the formula for the complex cisplatin.

Summary

- Amino acids are aminocarboxylic acids; in acid solution they form cations and in alkaline solution they form anions.
- Proteins are condensation polymers of amino acids; the amino acids are linked by peptide groups.
- Peptides can be hydrolysed using acid or base.
- Proteins have primary, secondary and tertiary structures.
- DNA is a polymer of nucleotides; nucleotides are composed of a 2-deoxyribose sugar bonded to a phosphate group and a base.
- The bases present in DNA are guanine, cytosine, adenine and thymine.
- Cisplatin is a complex that acts as an anticancer drug.

Organic synthesis

One functional group in organic chemistry can be converted into a different functional group. These conversions are the basis of organic synthesis:

- They often occur in a series of steps but chemistry tries to limit the number of steps as there is waste and potentially loss of yield with each one.
- Auxiliary chemicals such as solvents tend to be avoided.
- Gas-phase reactions are preferred.
- Water is the solvent of choice if it can be used.
- Organic solvents are usually volatile and it is hard to control their release into the atmosphere.

Table 10 gives the organic reactions from the entire specification. You should be familiar with these reactions and be able to use them to plan an organic synthesis with up to four steps.

Table 10 The organic reactions summary

No	Example	Reagents	Conditions	Mechanism	Type of reaction
1	alkane $\rightarrow$ halogenoalkane $CH_4 + Cl_2 \rightarrow CH_3Cl + HCl$	Cl_2	UV light	Free-radical substitution	Substitution
2	alkene $\rightarrow$ halogenoalkane $C_2H_4 + HBr \rightarrow C_2H_5Br$	HX	N/A	Electrophilic addition	Addition
3	alkene $\rightarrow$ dihaloalkane $C_2H_4 + Br_2 \rightarrow CH_2BrCH_2Br$	Halogen, e.g. Br_2	N/A	Electrophilic addition	Addition
4	alkene $\rightarrow$ alcohol $C_2H_4 + H_2O \rightarrow C_2H_5OH$ 1°, 2° and 3° alcohols can be produced depending on the position of the C=C and the actual alkene used	H_2O	Concentrated H_2SO_4 or concentrated H_3PO_4 300°C 7 MPa	Electrophilic addition	Addition followed by hydrolysis
5	1° halogenoalkane $\rightarrow$ 1° amine $C_2H_5Br + NH_3 \rightarrow C_2H_5NH_2 + HBr$	Concentrated NH_3	Excess NH_3 dissolved in ethanol	Nucleophilic substitution	Substitution
6	1° halogenoalkane $\rightarrow$ 2° amine $C_2H_5Br + C_2H_5NH_2 \rightarrow (C_2H_5)_2NH + HBr$	1° amine	Amine dissolved in ethanol	Nucleophilic substitution	Substitution
7	1° halogenoalkane $\rightarrow$ 1° alcohol $C_2H_5Br + NaOH \rightarrow C_2H_5OH + NaBr$	NaOH(aq)	Heat under reflux	Nucleophilic substitution	Substitution
8	2° halogenoalkane $\rightarrow$ 2° alcohol $CH_3CHBrCH_3 + NaOH \rightarrow$ $\qquad CH_3CH(OH)CH_3 + NaBr$	NaOH(aq)	Heat under reflux	Nucleophilic substitution	Substitution
9	3° halogenoalkane $\rightarrow$ 3° alcohol $(CH_3)_3CBr + NaOH \rightarrow (CH_3)_3COH + NaBr$	NaOH(aq)	Heat under reflux	Nucleophilic substitution	Substitution
10	1° or 2° or 3° halogenoalkane $\rightarrow$ alkene $C_2H_5Br + KOH \rightarrow C_2H_4 + KBr + H_2O$ The position of the halogen atom and the actual halogenoalkane will give different alkenes or even a mixture	KOH (dissolved in ethanol)	Heat under reflux	Elimination	Elimination
11	1° halogenoalkane $\rightarrow$ nitrile $C_2H_5Br + KCN \rightarrow C_2H_5CN + KBr$	Potassium cyanide (dissolved in ethanol)		Nucleophilic substitution	Substitution
12	alcohol $\rightarrow$ alkene $C_2H_5OH \rightarrow C_2H_4 + H_2O$ 1° or 2° or 3° alcohols can undergo this reaction to form differing alkenes or a mixture of alkenes	Concentrated H_2SO_4 or concentrated H_3PO_4 (or Al_2O_3 catalyst)	170°C for acid dehydration	Elimination	Elimination Dehydration
13	alcohol $\rightarrow$ halogenoalkane $C_2H_5OH + HX \rightarrow C_2H_5X + H_2O$	HX (prepared in situ from NaX and conc H_2SO_4)	Heat under reflux	N/A	Substitution
14	1° alcohol $\rightarrow$ aldehyde $C_2H_5OH + [O] \rightarrow CH_3CHO + H_2O$	Acidified potassium dichromate(VI) solution	Heat and distil	N/A	Oxidation

$\rightarrow$

No	Example	Reagents	Conditions	Mechanism	Type of reaction
15	2° alcohol → ketone $CH_3CH(OH)CH_3 + [O] → CH_3COCH_3 + H_2O$	Acidified potassium dichromate(VI) solution	Heat under reflux	N/A	Oxidation
16	aldehyde → carboxylic acid $CH_3CHO + [O] → CH_3COOH$	Acidified potassium dichromate(VI) solution	Heat under reflux	N/A	Oxidation
17	aldehyde → 1° alcohol $CH_3CH_2CHO + 2[H] → CH_3CH_2CH_2OH$	$NaBH_4$	Aqueous solution	Nucleophilic addition	Reduction
18	ketone → 2° alcohol $CH_3COCH_3 + 2[H] → CH_3CH(OH)CH_3$	$NaBH_4$	Aqueous solution	Nucleophilic addition	Reduction
19	aldehyde/ketone → hydroxynitrile $CH_3CHO + HCN → CH_3CH(OH)CN$	KCN followed by dilute acid	N/A	Nucleophilic addition	Addition
20	carboxylic acid → aldehyde $CH_3COOH + 2[H] → CH_3CHO + H_2O$	$LiAlH_4$	In dry ether	Nucleophilic addition	Reduction
21	carboxylic acid → 1° alcohol $CH_3COOH + 4[H] → CH_3CH_2OH + H_2O$	$LiAlH_4$	In dry ether	Nucleophilic addition	Reduction
22	carboxylic acid → ester $CH_3COOH + C_2H_5OH → CH_3COOC_2H_5 + H_2O$	Alcohol	Concentrated sulfuric acid	Nucleophilic addition–elimination	Elimination or condensation
23	nitrile → 1° amine $C_2H_5CN + 4[H] → C_2H_5CH_2NH_2$	Lithal/$LiAlH_4$	In dry ether	Nucleophilic addition	Reduction
24	carboxylic acid → ammonium salt $CH_3COOH + NH_3 → CH_3COONH_4$	Ammonia solution	Room temperature	N/A	Neutralisation
25	carboxylic acid → sodium salt $CH_3COOH + NaOH → CH_3COONa + H_2O$	NaOH(aq) (or Na_2CO_3)	Room temperature	N/A	Neutralisation
26	acid chloride → carboxylic acid $CH_3COCl + H_2O → CH_3COOH + HCl$	H_2O	Room temperature	Nucleophilic addition–elimination	Hydrolysis
27	acid chloride → amide $CH_3COCl + NH_3 → CH_3CONH_2 + HCl$	Ammonia	Acid chloride added to concentrated ammonia	Nucleophilic addition–elimination	Substitution
28	acid chloride → ester $CH_3COCl + C_2H_5OH →$ $CH_3COOC_2H_5 + HCl$	Alcohol added to acid chloride	Room temperature	Nucleophilic addition–elimination	Esterification or elimination
29	ester → carboxylic acid $CH_3COOC_2H_5 + H_2O →$ $CH_3COOH + C_2H_5OH$	Dilute hydrochloric acid	Heat under reflux	N/A	Acid hydrolysis
30	ester → salt of carboxylic acid $CH_3COOC_2H_5 + NaOH →$ $CH_3COONa + C_2H_5OH$	Sodium hydroxide solution (or any alkali)	Heat under reflux	N/A	Base hydrolysis
31	acid anhydride → carboxylic acid $(CH_3CO)_2O + H_2O → 2CH_3COOH$	H_2O	Room temperature	N/A	Hydrolysis

→

No	Example	Reagents	Conditions	Mechanism	Type of reaction
32	acid anhydride → amide $(CH_3CO)_2O + NH_3 →$ $CH_3CONH_2 + CH_3COOH$	Concentrated NH_3	Room temperature	N/A	N/A
33	acid anhydride → ester $(CH_3CO)_2O + CH_3CH_2OH →$ $CH_3COOCH_2CH_3 + CH_3COOH$	Alcohol	Room temperature	N/A	Elimination or condensation
34	benzene → nitrobenzene $C_6H_6 + HNO_3 → C_6H_5NO_2 + H_2O$	Concentrated HNO_3 Concentrated H_2SO_4	Low temperature to prevent further nitration	Electrophilic substitution	Substitution
35	nitrobenzene → phenylamine $C_6H_5NO_2 + 6[H] → C_6H_5NH_2 + 2H_2O$	Sn HCl	Heat under reflux and add NaOH(aq) to liberate the free amine	N/A	Reduction
36	benzene → phenylethanone $C_6H_6 + CH_3COCl → C_6H_5COCH_3 + HCl$ $C_6H_6 + (CH_3CO)_2O →$ $C_6H_5COCH_3 + CH_3COOH$	CH_3COCl or $(CH_3CO)_2O$	$AlCl_3$ catalyst with CH_3COCl	Electrophilic substitution	Substitution

Worked example

$CH_3CH_2CH_2Br$ is converted to $CH_3CH_2CH_2CH_2NH_2$ in a two-step process.

a Identify the intermediate.

b Give the reagent and condition for both steps.

Answer

a $CH_3CH_2CH_2CN$ or butanenitrile

b Step 1: KCN/potassium cyanide dissolved in ethanol
   Step 2: lithal/$LiAlH_4$ in dry ether

Nuclear magnetic resonance spectroscopy

The nuclei of ^1H and ^{13}C carbon atoms have magnetic properties. The environment in which these nuclei are found can be determined from nuclear magnetic resonance (NMR) spectroscopy and the information used to elucidate structure or parts of structure of an organic molecule. The nuclei act as little magnets and align their poles with an external magnetic field.

There are some key points about NMR spectroscopy:

- The comparative energy required to flip the magnetic field of the nuclei is measured as a chemical shift, which is represented by the symbol δ.
- Chemical shift is measured in parts per million (ppm).
- All chemical shift values are measured relative to a standard, which is tetramethylsilane (TMS).
- The chemical shift scale goes from 0 ppm for TMS on the right and runs to the left.
- Typical chemical shift values for ^1H nuclei are between 0.5 and 12.0, depending on the environment.

Exam tip

The key to this type of question is that the carbon chain is getting longer. This usually involves the production of a nitrile from a halogenoalkane. The nitrile can then be reduced to the amine. These are reactions 11 and 23 in Table 10.

Exam tip

Try to convert ethene to ethanal in a three step-process. The reactions required for a three-step process are 2, 7 and 14 in Table 10. It could be done in two steps using 4 and 14, but a three-stage process is required here.

Content Guidance

- Typical chemical shift values for ^{13}C nuclei are between 5 and 220, again depending on the chemical environment.
- All nuclei in the same chemical environment will have the same chemical shift.
- When ^{1}H nuclei or ^{13}C nuclei are found close to electronegative atoms such as oxygen the chemical shift values for these nuclei will be larger. They are described as being deshielded.
- The solvent used in NMR spectroscopy is tetrachloromethane, CCl_4, or any deuterium-containing solvent such as $CDCl_3$, CD_2Cl_2 or C_2D_6, where D represents deuterium.
- Deuterated solvents do not give a signal on ^{1}H NMR spectroscopy but do give a signal on ^{13}C NMR spectroscopy; this can be removed by computer software.
- An integration trace is often produced with an ^{1}H NMR spectrum, which integrates the areas under the peaks. The ratios are the same as the ratio of the number of hydrogen nuclei in each environment.
- For high-resolution ^{1}H NMR spectroscopy the signal for an environment is split into a series of peaks depending on the number of hydrogen nuclei bonded to adjacent carbon atoms.
- The $(n + 1)$ rule, where n is the number of hydrogen nuclei, gives the number of peaks into which the signal for the neighbouring environment will be split.

Knowledge check 36

Name a suitable intermediate in a two-step process for the conversion of propene to propan-2-ol.

Knowledge check 37

What is the standard used in NMR spectroscopy?

Worked example

^{1}H NMR spectroscopy

The ^{1}H NMR spectrum in Figure 83 is for a compound with molecular formula $C_4H_8O_2$. What features of this molecule can be elucidated from its formula and the spectrum?

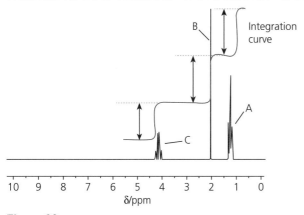

Figure 83

Answer

From the formula it is likely that the compound is an ester or a carboxylic acid. It is not a diol because there would be more hydrogen atoms; it could contain a C=O and an O–H group separately, but this would lead to two single peaks — one for a CH_3 beside a C=O and another for the O–H.

The possible structures for a carboxylic acid or an ester with this molecular formula are shown in Figure 84.

→

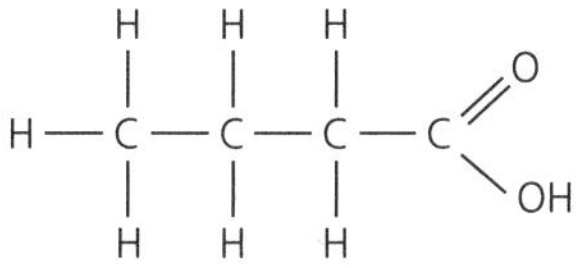

Butanoic acid

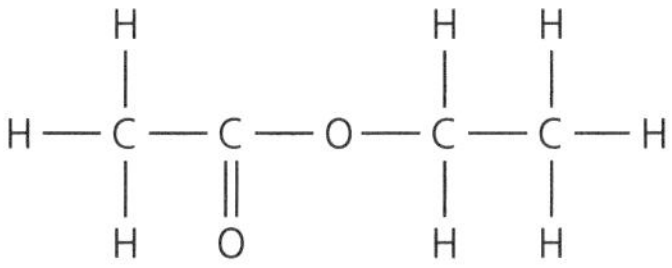

Ethyl ethanoate

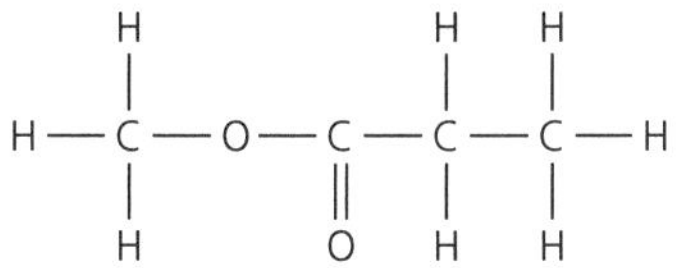

Methyl propanoate

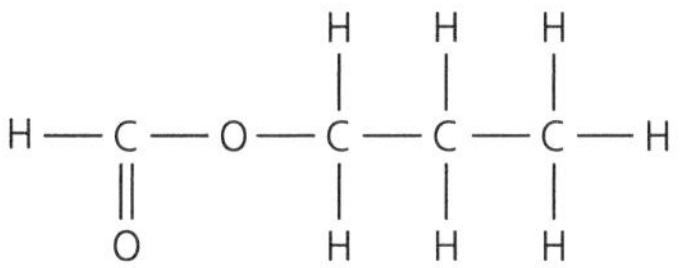

Propyl methanoate

Figure 84

There are three environments of chemically equivalent ^{1}H nuclei in this molecule. This is because there are three distinct peaks labelled A, B and C.

The molecule cannot be butanoic acid — it would have four different environments because there are four distinct environments for ^{1}H nuclei in the molecule. These are shown in Figure 85 (each environment is given a different colour).

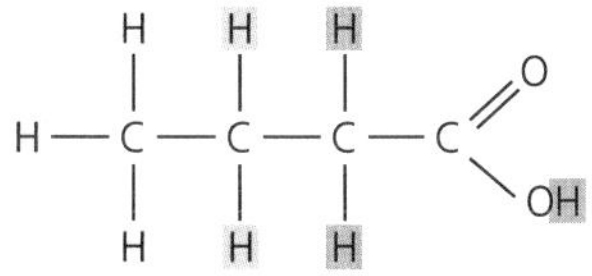

Figure 85

Similarly, propyl methanoate would have four different environments of ^{1}H nuclei.

The two options remaining are ethyl ethanoate and methyl propanoate. Both have three environments of chemically equivalent ^{1}H nuclei. These are shown in Figure 86.

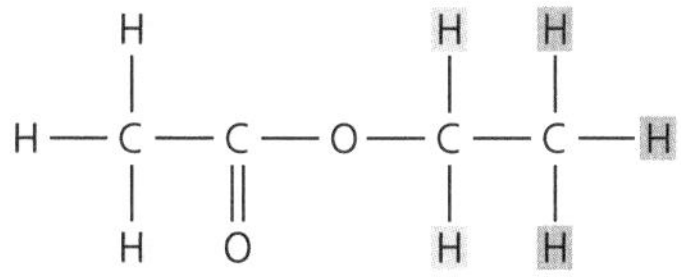

Ethyl ethanoate

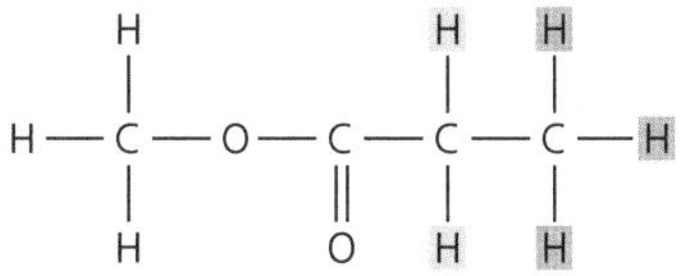

Methyl propanoate

Figure 86

The integration trace would suggest three ^{1}H nuclei in the environment at $\delta = 1.2$, three ^{1}H nuclei in the environment at $\delta = 2.1$ and two ^{1}H nuclei in the environment at $\delta = 4.1$. This is calculated from the actual ratio of the distance dropped by the integration curve over the peaks. Data may also be given in a table format, such as Table 11.

Table 11

δ/ppm	Peak integration	Spin–spin splitting pattern
1.2	3	Triplet
2.1	3	Singlet
4.1	2	Quartet

Both the isomers above have a 3:3:2 ratio of ^1H nuclei.

The spin–spin splitting pattern may also be considered:

- In ethyl ethanoate the yellow environment has no ^1H nuclei bonded to adjacent carbon atoms, so will appear as a single peak (called a singlet).
- The green environment has three ^1H nuclei bonded to an adjacent carbon atom, so will appear as four peaks (a quartet).
- The blue environment has two ^1H nuclei bonded to adjacent carbon atoms, so will appear as three peaks (a triplet).

In methyl propanoate the same pattern of a singlet, a quartet and a triplet appears. The chemical shift of the singlet and the quartet are the important features. In methyl propanoate the triplet is caused by the CH_3 group that is directly adjacent to an oxygen atom, which would mean it should be most deshielded. In ethyl ethanoate the quartet should be the most deshielded because the CH_2 group (which causes the quartet) is directly adjacent to an oxygen atom, so it should be the most deshielded.

The quartet is the most deshielded in the ^1H NMR spectrum, so this would suggest that the spectrum is that of ethyl ethanoate.

The ^1H NMR chemical shifts that are provided in the data booklet are shown in Table 12.

Table 12 ^1H NMR chemical shift data

Type of proton	δ/ppm	Type of proton	δ/ppm
ROH	0.5–5.0	R—O—C—H	3.1–3.9
RCH$_3$	0.7–1.2	RCH$_2$Cl or Br	3.1–4.2
RNH$_2$	1.0–4.5	R—C(=O)—O—C—H	3.7–4.1
R$_2$CH$_2$	1.2–1.4	R\C=C/H	4.5–6.0
R$_3$CH	1.4–1.6	R—C(=O)—H	9.0–10.0
R—C(=O)—C—H	2.1–2.6	R—C(=O)—O—H	10.0–12.0

Exam tip

A peak that is split into two is called a doublet. A peak split into more than four is often called a multiplet, but can be called a quintet, sextet, septet, octet etc. These are more rare.

The chemical shifts in the ^1H spectrum are 1.2, 2.1 and 4.1.

The peak at 1.2 corresponds to RCH_3 between 0.7 and 1.2; the peak at 2.1 corresponds to $RCOCH$ between 2.1 and 2.6; the peak at 4.1 corresponds to the peak for $RCOOCH$ between 3.7 and 4.1. This would confirm that it is ethyl ethanoate. If it were methyl propanoate, the singlet would be between 3.7 and 4.1.

Knowledge check 38

What is the spin–spin splitting pattern observed in a ^1H NMR spectrum for chloroethane?

Worked example

^{13}C NMR spectroscopy

Two organic compounds, labelled A and B, are shown in Figure 87.

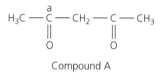

Compound A

Compound B

Figure 87

a State the number of peaks in the ^{13}C NMR spectrum of compound A and compound B.

b Give the IUPAC name for compound A.

c Suggest a value for δ in the ^{13}C NMR spectrum for the carbon atom labelled a.

Answer

a Examining compound A. In ^{13}C NMR carbon atoms in the same environment have the same chemical shift or δ value. Symmetry plays an important part, as does proximity to other groups or atoms.

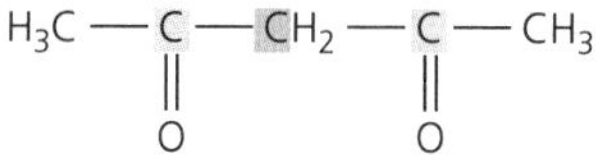

$$H_3C - C - CH_2 - C - CH_3$$
$$\qquad \| \qquad \qquad \|$$
$$\qquad O \qquad \qquad O$$

Figure 88

In Figure 88 the carbon atoms labelled in yellow have the same chemical shift in ^{13}C NMR. This is because of the symmetry of the molecule — they are in the same position relative to the C=O and also the CH_2 in the middle of the molecule. The carbon atoms labelled in green also have the same chemical shift as they are in the same position relative to the CH_3 groups and the CH_2 in the middle. The carbon atom labelled in blue has a separate chemical shift. There are *three* peaks on the ^{13}C NMR spectrum for compound A.

Examining compound B. Watch out for symmetry with aromatic compounds. Look for the same proximity to the groups bonded to the benzene ring.

$\rightarrow$

The two carbon atoms labelled pink in Figure 89 have the same chemical shift because they are equidistant from the CH_3 group and the Cl group. Similarly for the two labelled in green. The yellow-, red- and blue-labelled carbons all have separate chemical shift values, so there are *five* peaks in the ^{13}C NMR spectrum of compound B.

b A is pentane-2,4-dione.

c The ^{13}C NMR chemical shifts that are provided in the data booklet are shown in Table 13. You will see RCOCH indicated as having a chemical shift of 190–220 ppm.

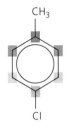

Figure 89

Table 13 ^{13}C NMR chemical shift data

Type of carbon	δ/ppm	Type of carbon	δ/ppm
—C—C—	5–40	C=C	90–150
R—C—Cl or Br	10–70	R—C≡N	110–125
R—C—C— (with O)	20–50	(benzene ring)	110–160
R—C—N	25–60	R—C— Esters or acids (with O)	160–185
—C—O— Alcohols, ethers or esters	50–90	R—C— Aldehydes or ketones (with O)	190–220

For more complex molecules with two or more benzene rings, the carbon atoms in the ring have the same chemical shift if they are equidistant from the groups linking them or the groups attached to them. The molecule in Figure 90 is phenolphthalein. There are 12 peaks on the ^{13}C NMR spectrum for this molecule. They are colour coded here. See if you can work out why some would have the same δ value.

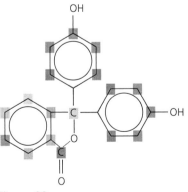

Figure 90

Chromatography

Chromatography is a separation technique for separating mixtures. It is used as an analytical technique in which the components of a mixture may be identified.

In chromatography there is a stationary phase and a mobile phase. The mobile phase is a liquid or a gas and the stationary phase is often a solid or a liquid held on a solid support.

The substances to be separated are dissolved in the mobile phase and, as the mobile phase moves through the stationary phase, the substances partition themselves between the two phases. Those substances that stay mainly in the mobile phase move fastest and those that are held by the stationary phase move least.

There are different types of chromatography but the ones we consider here are:

- thin layer chromatography (TLC)
- column chromatography
- gas chromatography

Each type has a mobile phase and a stationary phase. Table 14 details the phases.

Table 14

Type of chromatography	Mobile phase	Stationary phase
Thin-layer	Liquid solvent (e.g. water or organic solvents)	Solid silica gel paste on a microscope slide or plastic plate (or sometimes referred to as the solvent or water in the gel)
Column	Liquid solvent (e.g. water or organic solvents)	Solid silica gel
Gas	Inert carrier gas (e.g. N_2, Ne)	Microscopic film of liquid on a solid support

Thin-layer chromatography

Thin-layer chromatography (TLC) uses a plate of silica gel.

1 A pencil line is drawn about 1 cm from the bottom of the plate with a × in the middle of the line. This is the origin.

2 The sample to be separated (or identified) is dissolved in the solvent (or mixture of solvent) and spotted onto the × and allowed to dry.

3 The plate is then placed upright in a small volume of the solvent at the bottom of a beaker. The spot and pencil line should not be immersed in the solvent.

4 As the solvent is drawn up the plate the substances in the sample will move with the solvent.

5 When the solvent is close to the top, the plate should be removed and the position of the solvent front marked using a pencil line.

6 The spots can be developed using a chemical developing agent (such as ninhydrin for amino acids) or viewed under ultraviolet light. Under the ultraviolet light the spots can be drawn around using a pencil.

Figure 91 shows a typical TLC plate with some spots labelled A to E. The position of the origin and solvent front are also shown.

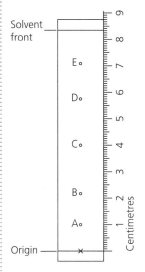

Figure 91

$$\text{retardation factor } (R_f) = \frac{\text{distance moved by spot}}{\text{distance moved by solvent}}$$

Spot E is the most soluble in the solvent (mobile phase) and least held back by the stationary phase, so moves furthest on the TLC plate. For E:

$$R_f = \frac{7.2}{8.3} = 0.867$$

The silica gel holds polar substances back whereas non-polar substances travel rapidly up a TLC plate.

Under similar conditions the R_f value can be used to identify the substance.

> **Required practical 12**
>
> You will be required to carry out a separation of a mixture of species using thin-layer chromatography.

Column chromatography

Column chromatography uses a column packed with solid silica gel. The solvent is added to the top of the column continually. The column should never dry out.

The substances to be separated are added at the top and as for TLC will move through the silica gel at different rates depending on their solubility in the solvent and their retention by the solid.

Figure 92 summarises column chromatography.

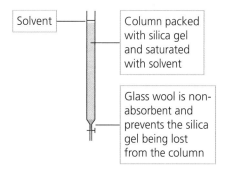

Figure 92 Column chromatography

The substances are eluted from the bottom of the column as they separate. Coloured substances are easily observed as they elute from the column. Colourless substances are not visible, so samples of the eluent (fluid eluted from the column) are taken and analysed. As with TLC polar substances are retained by the silica gel and move more slowly through the column.

Gas chromatography

Gas chromatography is also called gas–liquid chromatography because the stationary phase is a liquid and the mobile phase is a gas. Substances are separated by partition between the liquid stationary phase and the mobile gas phase. Inert gases such as argon or nitrogen are used as carrier gases. The column is coiled to make it compact and is held in an oven to maintain the substances as gases in the column. The process is shown in Figure 93.

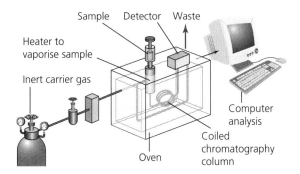

Figure 93 Gas chromatography

The computer monitors the substances eluted from the column and produces a trace (Figure 94).

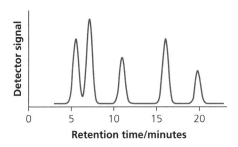

Figure 94

The retention time is the time the substances are held in the column before being eluted. The computer also integrates the area under each peak, which is proportional to the amount of substance present. A pure substance will have the same retention time in gas chromatography.

Gas chromatography can also be linked to mass spectrometry. Separated components are sent to the mass spectrometer for identification.

Knowledge check 40

What is meant by R_f?

Content Guidance

Summary

- Organic synthesis routes should contain the fewest possible steps to minimise loss and waste.
- The types of reaction, reagents, conditions and any mechanisms should be known for every organic reaction.
- ^1H NMR spectroscopy shows different environments of chemically equivalent ^1H nuclei within molecules.
- ^1H nuclei in the same chemical environment occur at the same chemical shift (δ) on an ^1H NMR spectrum.
- The areas under the peaks in an ^1H NMR spectrum are called integrations and are equivalent to the ratio of the number of ^1H nuclei in each environment.
- ^1H nuclei bonded to adjacent carbon atoms cause spin–spin splitting; the spin–spin splitting pattern is $n + 1$ peaks for n ^1H nuclei bonded to adjacent carbon atoms.
- ^{13}C NMR spectroscopy shows different environments of chemically equivalent ^{13}C nuclei. Chemically equivalent ^{13}C nuclei occur at the same chemical shift (δ).
- All types of chromatography have a mobile phase and a stationary phase.
- Components in chromatography are separated based on their affinity for the mobile or stationary phase.
- The R_f value is the distance moved by the spot divided by the distance moved by the solvent in TLC.
- In column chromatography, substances are eluted from a silica gel-based column. Polar substances are retained longer by the polar silica gel column.
- In gas chromatography, a plot of detector signal against retention time is obtained; a peak shows the presence of a substance.
- Integration of the trace gives a measure of the relative concentrations of substances in the mixture.
- Substances can be identified by their R_f values and retention times, as long as these are the same under the same conditions.

Questions & Answers

The examinations

The A-level examination consists of three examinations of 2 hours each. Papers 1 and 2 comprise short and long structured questions with a total mark of 105 each. Paper 3 covers the entire specification and contains practical and data analysis questions as well as multiple-choice questions worth 30 marks. For each multiple-choice question there is one correct answer and at least one clear distractor.

A-level paper 1 contains topics of physical chemistry (except 3.1.5 and 3.1.9) and includes inorganic chemistry (3.2). A-level paper 2 covers organic chemistry (3.3) and physical chemistry 3.1.2 – 3.1.6 and 3.1.9. A-level paper 3 covers the entire specification.

The questions in this section are on inorganic and organic chemistry sections 3.2.4–3.2.6 and 3.3.7–3.3.16. There are some questions linked to other parts of the specification. All other chemistry is covered in the first three student guides in this series.

About this section

This section contains a mix of multiple-choice and structured questions similar to those you can expect to find in the A-level papers.

Answers to the questions are followed by comments, preceded by the icon **e**. Try the questions first to see how you get on and then check the answers and comments.

General tips

- Be accurate with your learning at this level — examiners will penalise incorrect wording.
- At least 20% of the marks in assessments for chemistry will require the use of mathematical skills. For any calculation, always follow it through to the end even if you feel you have made a mistake — there are marks for the correct method even if the final answer is incorrect.
- Always attempt to answer a multiple-choice question even if it is a guess (you have a 25% chance of getting it right).

The uniform mark you receive for each of paper 1 and paper 2 will be out of 105. The uniform mark for paper 3 is out of 90. The total marks for A-level Chemistry are 300.

■ Properties of period 3 elements and their oxides

Question 1

Which of the following oxides of period 3 elements produces a solution with the highest pH?

A Na_2O **B** MgO **C** P_4O_{10} **D** SO_3 (1 mark)

> Answer is A ✓

ⓔ The metal oxides are basic or amphoteric. Basic oxides react with acids but some may react with water to form solutions with a pH >7. Sodium oxide reacts with water to form sodium hydroxide solution, which has a pH of around 12–14. Magnesium oxide has a very strong ionic lattice and therefore a very low solubility in water. The pH of the resulting solution is around 9. The oxides of non-metals such as P_4O_{10} and SO_3 are acidic and would dissolve in water to give pH values between 0 and 2.

Question 2

Write equations for the following reactions of oxides:

(a) sodium oxide with water   (1 mark)

(b) potassium oxide with phosphorus(v) oxide   (2 marks)

(c) magnesium hydroxide with sulfur(vi) oxide   (2 marks)

> **(a)** $Na_2O + H_2O \rightarrow 2NaOH$ ✓
>
> **(b)** $6K_2O + P_4O_{10} \rightarrow 4K_3PO_4$ ✓✓
>
> **(c)** $Mg(OH)_2 + SO_3 \rightarrow MgSO_4 + H_2O$ ✓✓

ⓔ These reactions are often asked for in the exam. Part (a) involves recalling that oxides of metals form alkaline solutions if they react with water. Make sure that you know the anion that is formed from the acidic oxide or its acid. You can then write the formula of the salt as the product of the reaction. Any hydrogen in a hydroxide or an acid on the left can be balanced using water on the right. The same applies to elements in other periods. K_2O is used here in place of Na_2O but the same chemistry applies.

▪ Transition metals and ions in aqueous solution

Question 1

Which of the following is correct? (1 mark)

	Complex	Coordination number	Shape
A	$[Co(H_2O)_6]^{2+}$	6	Octahedral
B	$[CoCl_4]^{2-}$	4	Square planar
C	$[Pt(NH_3)_2Cl_2]$	2	Tetrahedral
D	$[Ag(NH_3)_2]^+$	2	Bent

Answer is A ✓

🅔 The shapes of complexes are linear, octahedral, square planar or tetrahedral. 6-coordination complexes are octahedral. 4-coordination complexes are either square planar or tetrahedral and 2-coordination complexes are linear. $[CoCl_4]^{2-}$ is tetrahedral, $[Pt(NH_3)_2Cl_2]$ is square planar and $[Ag(NH_3)_2]^+$ is linear.

Question 2

Which of the following is the energy associated with a wavelength of 550 nm?
($c = 3.00 \times 10^8 \, m \, s^{-1}$; $h = 6.63 \times 10^{-34} \, J \, s$)

A $1.21 \times 10^{-36} \, J$ C $1.21 \times 10^{-27} \, J$

B $3.62 \times 10^{-28} \, J$ D $3.62 \times 10^{-19} \, J$ (1 mark)

Answer is D ✓

🅔 The calculation of energy from wavelength involves the use of $\Delta E = hc/\lambda$, where h is Planck's constant, c is the speed of light and λ is the wavelength. Wavelength should be converted to metres by multiplying by 10^{-9} before being used in the calculation. The other answers are obtained by omitting a step: A is obtained by not using c and not converting wavelength to m; B is obtained by using wavelength in nm not m (this can be a common mistake); C is obtained by omitting to use c.

Question 3

For the reaction $[Ni(NH_3)_6]^{2+} + EDTA^{4-} \rightarrow [NiEDTA]^{2-} + 6NH_3$, explain why ΔG is negative.

(4 marks)

> Same number of coordinate bonds broken and formed/ΔH = approximately 0. ✓
>
> Two particles in solution changes to seven particles. ✓
>
> Entropy increases. ✓
>
> $\Delta G = \Delta H - T\Delta S$ so ΔG is negative as $\Delta H = 0$ and ΔS is positive. ✓

ⓔ Both complexes in the reaction have a coordination number of 6 as $EDTA^{4-}$ is a hexadentate ligand. As the same number of coordinate bonds are broken and formed, ΔH for the reaction is approximately zero. It is not the important factor. The two ions on the left-hand side of the equation count as two particles and the complex and $6NH_3$ on the right count as seven particles. More particles in the solution means that there is an increase in disorder so an increase in entropy: $\Delta S > 0$. In the expression $\Delta G = \Delta H - T\Delta S$, if ΔH = approximately 0 and ΔS is positive, ΔG will be negative.

Question 4

2.48 g of an impure sample of iron(II) ethanedioate, FeC_2O_4, is dissolved in water and the volume made up to 250 cm^3 using deionised water. 25.0 cm^3 of this solution required 11.4 cm^3 of a 0.0540 mol dm^{-3} potassium manganate(VII) solution. Calculate the percentage purity of the sample of iron(II) ethanedioate. Give your answer to three significant figures.

(5 marks)

> moles of $MnO_4^- = \dfrac{11.4 \times 0.0540}{1000} = 6.156 \times 10^{-4}$ mol ✓
>
> ratio of $MnO_4^-:FeC_2O_4 = 3:5$
>
> moles of FeC_2O_4 in 25 $cm^3 = \dfrac{6.156 \times 10^{-4}}{3} \times 5 = 1.026 \times 10^{-3}$ mol ✓
>
> moles of FeC_2O_4 in 250 $cm^3 = 1.026 \times 10^{-3} \times 10 = 0.01026$ mol ✓
>
> mass of $FeC_2O_4 = 0.01026 \times 143.8 = 1.475$ g ✓
>
> % purity of same $= \dfrac{1.475}{2.48} \times 100 = 59.5\%$ ✓

ⓔ The ratio is all important in this question as a 3:5 ratio is unusual but you should remember these ratios as they will make your life a lot easier with these questions. The ratio of $Fe^{2+}:MnO_4^-$ is 5:1; the ratio of $C_2O_4^{2-}:MnO_4^-$ is 5:2 so combining these ratios gives 5:3 for the ratio of $FeC_2O_4: MnO_4^-$ because the FeC_2O_4 contains both the Fe^{2+} ion and the $C_2O_4^{2-}$ ion. Both of these ions react with MnO_4^-.

Question 5

The table below gives some electrode potentials.

Reaction	$E^{\ominus}/V$
$Ag^+(aq) + e^- \rightarrow Ag(s)$	+0.80
$Fe^{3+}(aq) + e^- \rightarrow Fe^{2+}(aq)$	+0.77
$Cu^{2+}(aq) + 2e^- \rightarrow Cu(s)$	+0.34
$Cd^{2+}(aq) + 2e^- \rightarrow Cd(s)$	−0.40
$Fe^{2+}(aq) + 2e^- \rightarrow Fe(s)$	−0.44
$Zn^{2+}(aq) + 2e^- \rightarrow Zn(s)$	−0.76
$Mn^{2+}(aq) + 2e^- \rightarrow Mn(s)$	−1.19

(a) Which species is the strongest reducing agent? (1 mark)

(b) Write the conventional cell representation for the cell formed between copper and cadmium electrodes. (2 marks)

(c) Calculate the EMF of this cell. (1 mark)

(d) What is observed when sodium hydroxide solution is added to a solution containing iron(II) ions? Write an equation for this reaction. (3 marks)

(e) Sodium carbonate solution is added to a solution containing iron(III) ions.

    (i) What is observed? (2 marks)

    (ii) Write an equation for this reaction (2 marks)

(a) Mn/manganese ✓

ℯ Manganese is the most easily oxidised so it is the strongest reducing agent. Change the sign of the electrode potentials so they represent oxidation reactions and $Mn \rightarrow Mn^{2+} + 2e^-$ has the largest positive value. The strongest oxidising agent in the table is silver (I) ions, Ag^+, because they have the highest positive value for a reduction and so are most easily reduced from the list in the table. Reducing agents should come from the right-hand side and oxidising agents from the left.

(b) $Cd|Cd^{2+}||Cu^{2+}|Cu$ ✓✓

ℯ The cell representation must have the oxidation on the left and the reduction on the right. Phase boundaries separate the components of the cell in different states or phases. The double line (||) represents the salt bridge. Each error with boundaries will lose a mark.

(c) EMF = +0.34 − (−0.40) = +0.74 V ✓

ⓔ The EMF can be calculated in several ways but here is it calculated as $E^{\ominus}_{rhs} - E^{\ominus}_{lhs}$. This is often the quickest way when cell representations are given or written.

(d) green ppt ✓

$[Fe(H_2O)_6]^{2+} + 2OH^- \rightarrow Fe(OH)_2(H_2O)_4 + 2H_2O$ ✓✓

ⓔ Both ammonia solution and sodium hydroxide solution produce a green precipitate with a solution containing iron(II) ions. It is important to write the equation forming $Fe(OH)_2(H_2O)_4$. These are the equations that are accepted in mark schemes.

(e) (i) brown ppt ✓
       bubbles of gas ✓

   (ii) $2[Fe(H_2O)_6]^{3+} + 3CO_3^{2-} \rightarrow 2Fe(OH)_3(H_2O)_3 + 3CO_2 + 3H_2O$ ✓✓

ⓔ The addition of carbonate ions to a solution containing 3+ ions results in the formation of the hydroxide precipitate and bubbles of gas (carbon dioxide) being released. Make sure that you can work out this equation. The iron components all have 2 as their balancing number and the others are all 3. The equation is the same for $[Al(H_3O)_6]^{3+}$, so once you know one equation you know the other. Observation wise, the only difference is that there is white ppt for the aluminium reaction.

■ Optical isomers, aldehydes and ketones, and carboxylic acids and derivatives

Question 1

Which one of the following molecules exhibits optical isomerism?   (1 mark)

A $CH_2(OH)CH_2COOH$ C $CH_3CH_2CH_2CHO$

B $CH_2CH(OH)CH_3$ D $CH_3CH(OH)COOH$

Answer is D ✓

ⓔ The molecule with a chiral centre (four different groups bonded to the same carbon atom) exhibits optical isomerism. Draw the structures of the organic molecules quickly in this type of question to determine the molecule with a chiral centre.

Question 2

Which one of the following compounds will react with Tollens' reagent?　　　　(1 mark)

A　$CH_3COCH_2CH_3$

C　CH_3CH_2CHO

B　$CH_3COOCH_2CH_3$

D　CH_3CH_2COOH

> Answer is C ✓

ⓔ Aldehydes react with Tollens' reagent, Fehling's solution and acidified potassium dichromate solution. Aldehydes undergo mild oxidation. Remember that primary and secondary alcohols also undergo mild oxidation and will react with acidified potassium dichromate solution. C is an aldehyde; you can recognise it from the CHO functional group. A is a ketone (CO group), B is an ester (COO group) and D is a carboxylic acid (COOH group).

Question 3

The following organic reaction scheme shows a series of reactions.

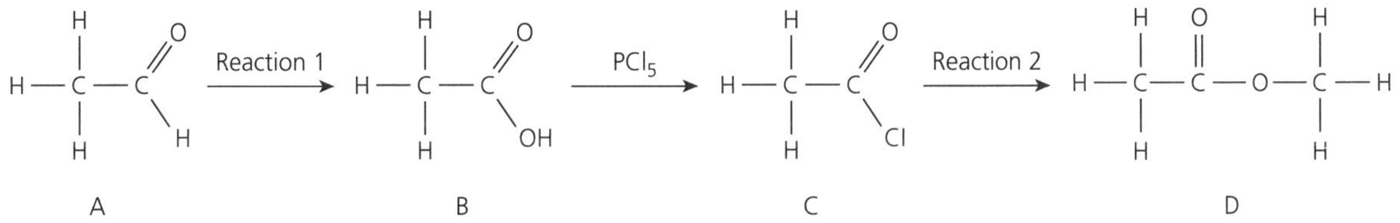

A　　　　　　　　　B　　　　　　　　　C　　　　　　　　　D

(a)　Name A, B, C and D.　　　　(4 marks)

(b)　State the reagents required for each reaction 1 and 2.　　　　(2 marks)

(c)　Write an equation for the reaction (2) to convert C to D.　　　　(1 mark)

> (a)　A is ethanal ✓
>
> 　　　B is ethanoic acid ✓
>
> 　　　C is ethanoyl chloride ✓
>
> 　　　D is methyl ethanoate ✓

ⓔ Look for the functional groups. Organic reactions focus on the change from one functional group to another. A has a CHO functional group and is an aldehyde. B has a COOH functional group and is a carboxylic acid. C has a COCl functional group and is an acyl chloride. D is an ester because it has the COO functional group. Organic reactions change one functional group into another functional group.

(b) Reaction 1 — any mild oxidising agent, for example acidified potassium dichromate solution ✓

Reaction 2 — methanol/CH_3OH ✓

ⓔ For reaction 1, Tollens' reagent and Fehling's solution are used as tests for aldehydes and ketones but acidified potassium dichromate would be the preferred reagent.

(c) $CH_3COCl + CH_3CH_2OH \rightarrow CH_3COOCH_2CH_3 + HCl$ ✓

ⓔ This type of question — composing organic compounds from different sections of the specification — is quite common.

Question 4

(a) **Carboxylic acids are weak acids. State the meaning of the term weak as applied to carboxylic acids.** (1 mark)

(b) **Write an equation for the reaction of propanoic acid with sodium carbonate and state two observations that would be made.** (2 marks)

(c) **Write an equation for the reaction of propanoic acid and ethanol, and name the ester produced.** (2 marks)

(a) They dissociate partially in solution. ✓

(b) $Na_2CO_3 + 2CH_3CH_2COOH \rightarrow 2CH_3CH_2COONa + H_2O + CO_2$ ✓

Heat/bubbles/solid disappears to form solution *(any 2)* ✓

ⓔ For part (b) ensure that you work out the formula of sodium carbonate correctly. Sodium is +1 and carbonate ion –2, hence the formula is Na_2CO_3. This is an acid and carbonate reaction — a salt, sodium propanoate, water and carbon dioxide are produced. Remember to balance your equation. For observations, remember that each time a gas — in this case carbon dioxide — is produced, the observation is bubbles. The solid carbonate will also disappear and form a solution.

(c) $C_2H_5OH + CH_3CH_2COOH \rightleftharpoons CH_3CH_2COOC_2H_5 + H_2O$ ✓

Ethyl propanoate ✓

ⓔ This is an esterification reaction. Remember that, when these substances react, water is removed. It is useful to draw this out structurally. Remember, too, that equilibrium arrows are needed. To name the ester, name the right-hand side and then the left as -anoate.

Question 5

(a) Name the compound CH_3CH_2CHO and state its functional group. (2 marks)

(b) Name and outline a mechanism for the reaction of CH_3CH_2CHO with HCN. (4 marks)

(c) What is meant by the term enantiomer? Draw the structure of each enantiomer formed in the reaction in (b). (3 marks)

(d) State and explain how you could distinguish between the two enantiomers. (2 marks)

(a) Propanal ✓

Carbonyl ✓

(b) Nucleophilic addition ✓

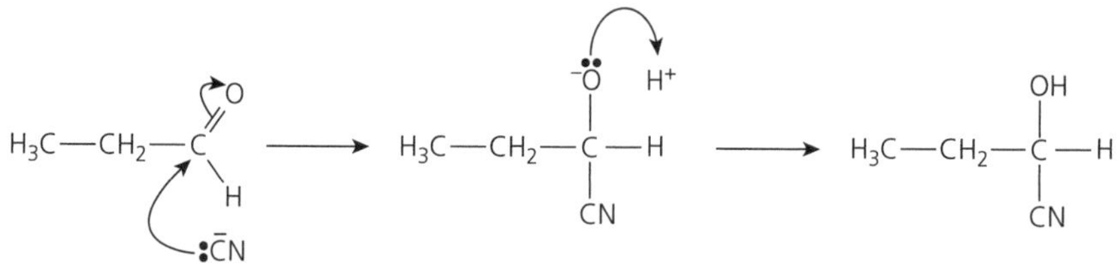

e This is nucleophilic addition and the nucleophile is CN⁻. Make sure you draw the curly arrows coming from the bond, and from the lone pair when forming a bond.

(c) Enantiomers exist as non-superimposable mirror images that differ in their effect on plane-polarised light. ✓

e When drawing enantiomers, first circle the four different groups on the molecule, then draw a tetrahedral shape and then place one group on each position. Finally, draw a mirror image for the second isomer.

(d) Plane-polarised light ✓ is rotated in opposite directions ✓.

Aromatic chemistry and amines

Question 1

Which one of the following is the mechanism by which ethanoic anhydride reacts with benzene? (1 mark)

A electrophilic addition   **C** nucleophilic addition

B electrophilic substitution   **D** nucleophilic substitution

> Answer is B ✓

> **e** Benzene undergoes substitution reactions rather than addition reactions because this preserves the stability of the delocalised system. The high electron density above and below the ring means that it is attacked by electrophiles.

Question 2

(a) Name compound A. (1 mark)

Compound A

(b) Name the reagents used to produce compound A from chlorobenzene and write an equation to show the reactive intermediate that is formed from these reagents. (2 marks)

(c) Name and outline the mechanism for the nitration of chlorobenzene to produce A. (3 marks)

(d) How would you experimentally distinguish between a sample of benzene and a sample of cyclohexene. Describe what you would see when the reagent is added to each compound and the test tube is shaken. (3 marks)

> **(a)** 4-chloronitrobenzene ✓

> **e** The compound is similar to nitrobenzene, $C_6H_5NO_2$, with which you are familiar. The chloro group is on position 4.

> **(b)** Concentrated nitric acid and concentrated sulfuric acid. ✓
>
> $HNO_3 + 2H_2SO_4 \rightarrow NO_2^+ + H_3O^+ + 2HSO_4^-$ ✓

> **e** The nitration of benzene to form nitrobenzene is a reaction that is on your specification. In this question you must realise that substance A has a nitro group, and so has been produced in a similar nitration reaction, which uses a nitrating mixture of concentrated nitric and sulfuric acid. The equation to show the reactive intermediate is one you must learn.

(c) Electrophilic substitution ✓

ℯ You need to have learned the mechanism for nitration of benzene. Apply this mechanism to the structure of chlorobenzene. Remember that the curly arrow must come from the benzene ring and the broken delocalised ring must be approximately centred on carbon.

(d) Bromine ✓

Benzene no reaction/colour remains ✓

In cyclohexene bromine is decolorised. ✓

ℯ Remember that benzene, unlike cyclohexene (an alkene), will not undergo addition reactions, and so it will not react with bromine. If bromine water colour is stated you would gain credit for red, yellow, orange, brown or any combination of these.

Question 3

Cetrimonium bromide has the formula $[CH_3(CH_2)_{15}N(CH_3)_3]^+Br^-$ and is found in many antiseptic creams.

(a) Name this type of compound. (1 mark)

(b) Give the reagent that must be added to $CH_3(CH_2)_{15}NH_2$ to make cetrimonium bromide and state the reaction conditions. (2 marks)

(c) Name the type of mechanism involved in this reaction. (1 mark)

(a) Quaternary ammonium salt/bromide ✓

(b) CH_3Br or bromomethane ✓

Excess CH_3Br or bromomethane ✓

(c) Nucleophilic substitution ✓

ℯ You must first realise that $CH_3(CH_2)_{15}NH_2$ is an amine and $[CH_3(CH_2)_{15}N(CH_3)_3]^+Br^-$ contains four alkyl groups, so it is a quaternary ammonium salt. To make these salts the amine must be treated with excess of the halogenoalkane, in this case bromomethane. The mechanism is nucleophilic substitution.

▪ Polymers

Question 1

Which of the following monomers would be suitable for making Kevlar?

A $HOCH_2CH_2OH$ and $HOOCC_6H_4COOH$

B CH_2CH_2 and CH_2CH_2

C $H_2NC_6H_4NH_2$ and $ClOCC_6H_4COCl$

D $H_2N(CH_2)_6NH_2$ and $HOOC(CH_2)_4COOH$ (1 mark)

> Answer is C ✓

ℯ Kevlar is a condensation polymer formed from benzene-1,4-dicarboxylic acid (or benzene-1,4-dioyl dichloride) and benzene-1,4-diamine. You could immediately rule out B as it would make polythene. A would make a PET. D would make nylon. Recognising the monomers in this format is important.

Question 2

Describe the primary, secondary and tertiary structure of a protein.   (6 marks)

> Primary structure is the sequence of amino acids ✓ joined by peptide links ✓.
>
> Secondary structure contains alpha- (α-) helices and beta- (β-) pleated sheets ✓ held together by hydrogen bonds between the C=O and N–H groups ✓.
>
> Tertiary structure is the final folding of the protein ✓ held together by ionic interactions/disulfide bridges/hydrophobic interactions ✓.

ℯ Details of the primary, secondary and tertiary structure of a protein are important to the overall 3D shape of the protein molecule. Make sure that you know the difference between these different levels of structure and could identify them from a diagram or description.

Question 3

Lysine and glutamic acid are two amino acids.

Lysine   Glutamic acid

(a) Give the IUPAC name of glutamic acid. (1 mark)

(b) Draw the structure to show the product when glutamic acid reacts with excess aqueous HCl. (1 mark)

(c) Draw the structure to show the product when glutamic acid reacts with excess aqueous NaOH. (1 mark)

(d) Draw the structure of a dipeptide formed from one molecule of lysine and one molecule of glutamic acid. (1 mark)

(a) 2-aminopentanedioic acid ✓

ⓔ Draw out the structure in full if it makes to easier to see it as pentanedioic acid with an amino (NH_2) group bonded to carbon 2.

(b)

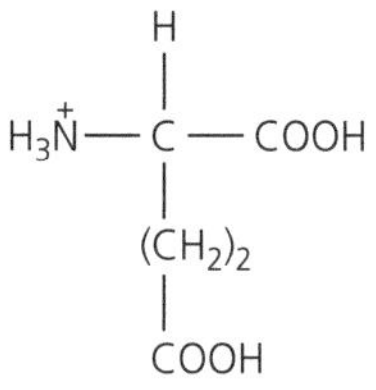

ⓔ In the presence of excess aqueous HCl, the NH_2 group is protonated. Had the question been about lysine, both NH_2 groups would be protonated. Make sure the + is beside or above the N atom. The COOH groups remain protonated.

(c)

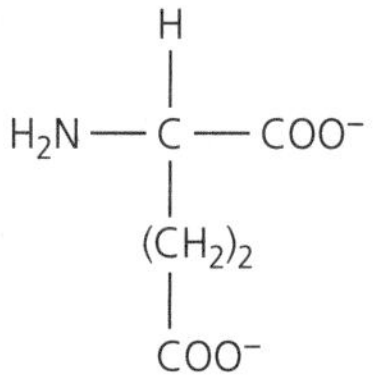

ⓔ In the presence of excess aqueous NaOH, the COOH groups are deprotonated. Make sure the – charge is beside the last O, as in COO^-.

(d)

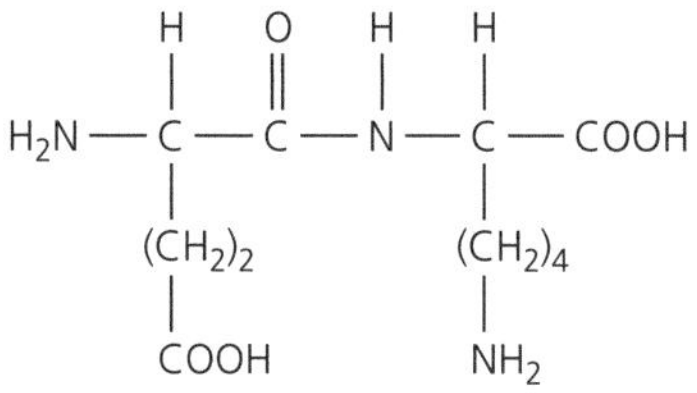

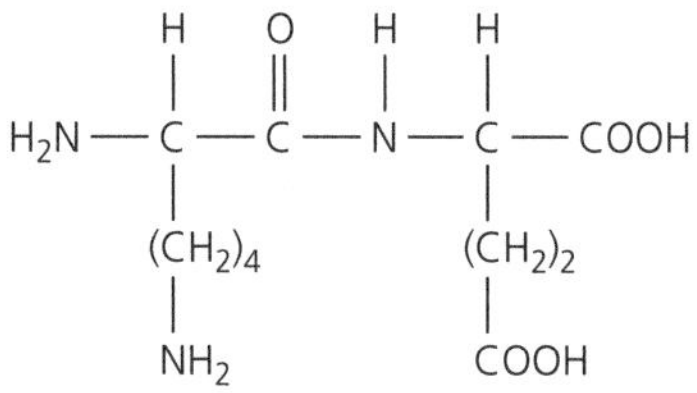

ⓔ There are two options here depending on which amino acid you draw first. The amino acids are connected by a peptide link, which may have the C=O group from the lysine or the glutamic acid. You could be expected to draw both.

Question 4

A dinucleotide is shown below.

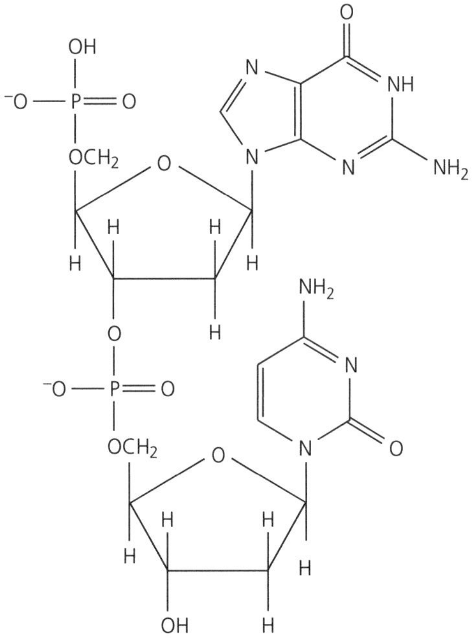

(a) Name bases A and B. (2 marks)

(b) Name the sugar in DNA. (1 mark)

(c) On base B, circle points that are used to form hydrogen bonds with a
complementary base. (1 mark)

(a) A is guanine. ✓   B is cytosine. ✓

ⓔ The bases are given in the data booklet with your examination. You should be
able to identify them from the structures given even when they are shown as part
of a portion of a strand of DNA or as a nucleotide.

(b) 2-deoxyribose ✓

(c)

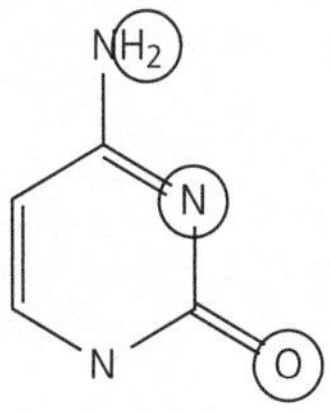

✓

ⓔ It is important to realise that GC base pairs have three hydrogen bonds
between the bases whereas AT base pairs only have two. So knowing this is
cytosine tells you that there are three points where it can form hydrogen bonds.

■ Organic synthesis, NMR spectroscopy and chromatography

Question 1

Which one of the following molecules would have two peaks in its ^{13}C NMR spectrum?

A $H_3C-CH_2-CH_2-COOH$

C $H_3C-CH_2-CH_2-CH_2-COOH$

B $HOOC-CH_2-CH_2-COOH$

D $HOOC-CH_2-CH_2-CH_2-COOH$ (1 mark)

> Answer is B ✓

ⓔ It is the symmetry of this molecule that gives it two peaks. A would have four peaks; C would have five peaks and D would have three peaks. The CH_2 groups in the middle of B have the same chemical shift due to the symmetry. The three peaks in D are caused by the COOH, two CH_2 beside the COOH and the central CH_2 group.

Question 2

C_2H_4 can be converted into $HOCH_2CH_2OH$, in a two-step process.

(a) Give the IUPAC name of C_2H_4 and $HOCH_2CH_2OH$. (2 marks)

(b) Identify the intermediate in the process and give the reagents and any conditions required for each step. (3 marks)

(c) Identify a polymer that the product can be used to manufacture. (1 mark)

> **(a)** C_2H_4 is ethene. ✓
>
> $HOCH_2CH_2OH$ is ethane-1,2-diol. ✓
>
> **(b)** 1,2-dibromoethane/CH_2BrCH_2Br ✓
>
> Bromine ✓
>
> Aqueous/NaOH/heat ✓
>
> **(c)** PET/PETE/poly(ethylene) terephthalate ✓

ⓔ It is important to be able to see synthetic routes in organic chemistry. Focus on the starting material and what you are aiming to achieve. In this example there are two OH groups on the product so these must have been substituted from another. Working backwards, it was probably a dihalogenoalkane (dibromoalkane is the most sensible as the Br groups are easily replaced with OH groups). The diol is used in the manufacture of PET using benzene-1,4-dicarboxylic acid.

Question 3

A dipeptide is shown below.

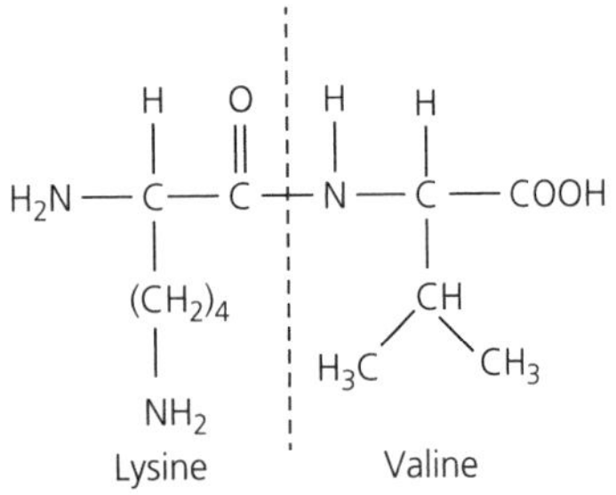

(a) The dipeptide is hydrolysed in acid conditions. The mixture is separated using column chromatography. The column is packed with a resin that acts as a polar stationary phase. Explain why lysine is eluted from the column after valine. (2 marks)

(b) Give the IUPAC name for valine. (1 mark)

(a) In acid conditions lysine has a greater positive charge than valine. ✓

Lysine has a greater affinity for the polar stationary phase. ✓

ⓔ Because lysine has NH_2 as part of its side chain, it has two positive groups when in acid solution. The stationary phase in column chromatography has a greater attraction of the more positively charged lysine. Valine is less charged and has a non-polar side chain so it is eluted before lysine. An acid side chain, such as aspartic acid, produced on alkaline hydrolysis would have a greater affinity for the polar stationary phase because of the two negative charges.

(b) 2-amino-3-methylbutanoic acid ✓

ⓔ Taking the amino acid out of a peptide and giving the IUPAC name for common amino acids are common questions. Break the peptide link and reform the amino acid. The carboxylic acid group takes priority over the amino group and the chain is numbered from the COOH group, with the C of the COOH group being carbon 1 in the chain. Amino comes before methyl alphabetically.

Question 4

Isomers of $C_5H_{10}O_2$ show differences in their ^1H NMR spectra. TMS is used as a standard in ^1H NMR spectroscopy.

(a) Apart from its lack of toxicity and inert nature, give two reasons why TMS is used as a standard. (2 marks)

(b) Draw the structure of a carboxylic acid isomer of $C_5H_{10}O_2$ that has two peaks in its ^1H NMR spectrum. (1 mark)

(c) Draw the structure of an unbranched ester isomer of $C_5H_{10}O_2$ that does not have a singlet in its ^1H NMR spectrum. Give the IUPAC name of the ester. (2 marks)

(a) TMS has 12 carbon atoms, all in the same environment, so a strong signal.

TMS peak is more shielded/more upfield than other peaks in organic molecules.

TMS has a low boiling point/is volatile/is easily removed. *(any two)* ✓✓

ⓔ The use of TMS as a standard is common along with the use of CCl_4 or $CDCl_3$ as a solvent. Make sure that you can draw the structure of TMS. It is based on silane, which is SiH_4 with the hydrogens replaced with four methyl groups. You should know that CCl_4 and $CDCl_3$ are used as solvents, because they do not contain ^1H. Also, CCl_4 is non-polar and therefore more useful for non-polar substances, whereas $CDCl_3$ is polar, so is often used to dissolve polar substances.

(b)

$$H_3C - \underset{\underset{CH_3}{|}}{\overset{\overset{CH_3}{|}}{C}} - COOH$$

✓

ⓔ Think carefully about the structure of the carboxylic acid that would only give two peaks. Often there will be symmetrical groups, such as the three CH_3 groups here.

(c) $H_3C - CH_2 - \underset{\underset{O}{\|}}{C} - O - CH_2 - CH_3$

✓

Ethyl propanoate ✓

ⓔ There are several unbranched esters, but many have a singlet. Sketch out the structures to see which has a CH_3 group beside the C=O or an H or CH_3 group beside the O, as these will form singlets in the ^1H NMR spectrum.

Knowledge check answers

1 Small higher charge ion/Mg^{2+} smaller than Na^+, so stronger ionic bonds in MgO.

2 $6CaO + P_4O_{10} \rightarrow 2Ca_3(PO_4)_2$

3 $MgO + SO_3 \rightarrow MgSO_4$
Magnesium sulfate(VI)

4 Sodium oxide

5 **a** $Fe = 1s^2\ 2s^2\ 2p^6\ 3s^2\ 3p^6\ 3d^6\ 4s^2$
　　b $Fe^{2+} = 1s^2\ 2s^2\ 2p^6\ 3s^2\ 3p^6\ 3d^6$
　　c $Fe^{3+} = 1s^2\ 2s^2\ 2p^6\ 3s^2\ 3p^6\ 3d^5$

6 **a** $[Co(H_2O)_6]^{2+}$, shape = octahedral, coordination number = 6
　　b $[Ag(NH_3)_2]^+$, shape = linear, coordination number = 2
　　c $[Pt(NH_3)_2Cl_2]$, shape = square planar, coordination number = 4

7 Five particles on the left and seven particles on the right, so it becomes more disordered.

8 Square planar

9 Anticancer drug

10 4.42×10^{-19} J

11 VO_2Cl

12 $Zn + 2V^{3+} \rightarrow Zn^{2+} + 2V^{2+}$

13 77.5%

14 Iron is a solid and nitrogen and hydrogen are gases. A heterogeneous catalyst is in a different phase/state from the reactants.

15 Green ppt

16 White ppt and bubbles of gas released

17 The second carbon from the left is a chiral centre. The third carbon from the left is also a chiral centre.

18 Butan-2-ol. Plane-polarised light is rotated in one direction by one optical isomer, and in a different direction by the other.

19 2-bromo-4-methylhexanal

20 Magnesium propanoate and water
$MgO + 2CH_3CH_2COOH \rightarrow (CH_3CH_2COO)_2Mg + H_2O$

21 Ethyl propanoate, $CH_3CH_2COOCH_2CH_3$

22 Biodiesel is a mixture of the following methyl esters:
$CH_3OOC(CH_2)_7CHCH(CH_2)_7CH_3$
$CH_3OOC(CH_2)_7CHCHCH_2CHCH(CH_2)_4CH_3$

23 $CH_3CH_2COCl + 2CH_3NH_2 \rightarrow CH_3CH_2CONHCH_3 + CH_3NH_3Cl$
N-methyl propanamide methylammonium chloride

24 Each carbon uses three of its outer electrons to form three sigma bonds, one to each carbon and one to each hydrogen, and each carbon has one involved in the delocalised ring.
$(4 \times 6) + 6$ electrons (one from each H) = 30

25 Dimethylamine

26 Phenylamine, ammonia, propylamine

27 Ethane-1,2-diol, benzene-1,4-dicarboxylic acid

28 PET, nylon-6,6, Kevlar

29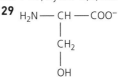

30 2-amino-4-methylpentanoic acid

31 Ninhydrin and ultraviolet

32 α-helix and β-pleated sheet

33 2-deoxyribose, $C_5H_{10}O_4$

34 Adenine, thymine, guanine, cytosine

35 $[Pt(NH_3)_2Cl_2]$

36 2-bromopropane or 2-chloropropane

37 Tetramethylsilane (TMS)

38 Triplet and quartet

39 Three

40 Retardation factor (or distance moved by spot/distance moved by solvent)

Index

Index